W9-CPE-507

MATHEMATICS TEACHING CASES

Number Sense and Operations in the Primary Grades

Hard to Teach and Hard to Learn?

Edited by

Carne Barnett-Clarke
Alma Ramirez

with

Debra Coggins and Susie Alldredge

Heinemann
Portsmouth, NH

Heinemann

A division of Reed Elsevier Inc.

361 Hanover Street

Portsmouth, NH 03801–3912

www.heinemann.com

Offices and agents throughout the world

© 2003 by WestEd

All rights reserved. No part of this book may be reproduced in any form or by any electronic or mechanical means, including information storage and retrieval systems, without permission in writing from the publisher, except by a reviewer, who may quote brief passages in a review. The cases that lead off each chapter may be photocopied for classroom or workshop use. This does not include the Facilitator's Guide Notes.

The authors and publisher wish to thank those who have generously given permission to reprint borrowed material:

A slightly different version of "Stop Sign or Equals Sign?" first appeared in the newsletter *Intersection,* April 2001. Adapted by permission of the ExxonMobil Foundation. Through its K–5 Mathematics Specialist Program, ExxonMobil supports projects to improve student learning in mathematics by strengthening teachers' grasp of content, pedagogy, and thoughtful assessment.

"Hundred Chart," "Count by Tens Chart," "Two Spinners," and "Is it Close Worksheet" are reprinted from *Number Power*™. Copyright © 1999 by Developmental Studies Center. Reprinted by permission of the Developmental Studies Center, 2000 Embarcadero, Suite 305, Oakland, CA 94606.

"Closer to 500 than 400," "Dollars and Cents Confusions," "Carry Two or Twenty?" and "How Many Can She Buy?" are adapted by permission of the Developmental Studies Center, 2000 Embarcadero, Suite 305, Oakland, CA 94606.

This publication is based on work supported by Stuart Foundation, the National Science Foundation (grant number 9550063), and the U.S. Department of Education, Office of Educational Research and Improvement (contract number 400-86-0009). Its contents do not necessarily reflect the views or policies of any of these agencies, nor does the mention of trade names, commercial products, or organizations imply endorsement by our sponsoring agencies.

Library of Congress Cataloging-in-Publication Data

Number sense and operations in the primary grades : hard to teach and hard to learn? / edited by Carne Barnett-Clarke, Alma Ramirez with Debra Coggins and Susie Alldredge.

    p. cm. — (Mathematics teaching cases)

  Includes bibliographical references.

  ISBN 0-325-00546-X (pbk. : alk. paper)

  1. Arithmetic—Study and teaching (Primary)—Case studies. 2. Number concept—Study and teaching (Primary)—Case studies. I. Barnett-Clarke, Carne. II. Ramirez, Alma. III. Series.

QA135.6 .N85 2003   2002151785

372.7'2—dc21

Editor: Victoria Merecki

Production: Lynne Reed

Cover design: Darci Mehall

Typesetter: Technologies 'N Typograhy, Inc.

Manufacturing: Steve Bernier

Printed in the United States of America on acid-free paper

07 06 05 04 03 RRD 1 2 3 4 5

Contents

Acknowledgments   v

Introduction   1

Using the Casebook and the Facilitator's Guide Notes   16

1 You Just Count the Extras   21

   Facilitator's Guide Notes   26

2 Word Problems First, Then Basic Facts   31

   Facilitator's Guide Notes   35

3 Everything in Its Place   39

   Facilitator's Guide Notes   43

4 Trading Games   47

   Facilitator's Guide Notes   52

5 Stop Sign or Equals Sign?   57

   Facilitator's Guide Notes   62

6 Daily Equations   67

   Facilitator's Guide Notes   74

7 Diana   79

   Facilitator's Guide Notes   84

8 Tallies and Coins 89

  Facilitator's Guide Notes 94

9 How Many More? 99

  Facilitator's Guide Notes 106

10 Why Is Subtraction More Difficult? 111

  Facilitator's Guide Notes 116

11 Tina 121

  Facilitator's Guide Notes 126

12 Is Number Sense Enough? 131

  Facilitator's Guide Notes 135

13 Closer to 500 than 400 141

  Facilitator's Guide Notes 148

14 Dollars and Cents Confusions 153

  Facilitator's Guide Notes 159

15 Carry Two or Twenty? 165

  Facilitator's Guide Notes 169

16 How Many Can She Buy? 175

  Facilitator's Guide Notes 181

Bibliography 184

Acknowledgments

For more than a decade, teachers in our project have devoted considerable time and creative energy to this project by writing, reviewing, and field-testing cases. We would especially like to express our gratitude to the primary grade teachers who contributed to this casebook. They have written candidly about their teaching to provide rich and authentic cases for discussion. We are also grateful to the other teachers who were involved in reviewing, field-testing, and revising cases. A special thank you to Norma Sakamoto who mentored case discussion groups as they field-tested these cases.

As staff members of the Mathematics Case Methods Project, we are proud to have provided opportunities for teachers to contribute to and learn from practice-based materials. This work would not have been possible without the long-term support and encouragement of Theodore Lobman, Ellen Hershey, and the board members at the Stuart Foundation. They provided us the privilege of pursuing ideas that were experimental and outside of the norm, beginning in 1989. In particular, we appreciate their determination to understand the aim of our work and their trust in our ability to accomplish our goals.

Some of the cases in this book were developed under the direction of Susie Alldredge at the Developmental Studies Center in Oakland, California. Her expertise and work with teachers on developmental and social issues adds a special dimension to this book. We also thank Eric Schaps, president of the Developmental Studies Center, for supporting this collaborative work. Debra Coggins, with her keen eye for detail and strong grounding in mathematics, also helped edit several of the cases and write facilitator's guide notes. She worked very hard to help us identify the important issues in the cases and to bring those issues to the forefront.

Also, we thank the ExxonMobil Foundation for their support of the K–5 Mathematics Specialist Program, which has touched the professional lives of the authors of several cases in this book.

We also want to acknowledge the careful and thoughtful editorial work of Joy Zimmerman with the WestEd communications staff. She suggested revisions for the cases that took into account the author's intentions, the content aims of the casebook, and the teachers who will be reading and discussing the cases. Finally, Maureen Metcalf, our project assistant, meticulously and thoughtfully carried out the preparation of the manuscript to be submitted for publication. She is especially appreciated for her helpful suggestions and willingness to put so much effort into making sure that all the details are complete and correct.

Commitment to the case discussion process is strong, as demonstrated by the hundreds of teachers, both nationally and internationally, who continue to discuss cases year after year and by those who take leadership roles to help others learn how to participate in the process. We invite you to learn how to start a case discussion group. To be placed on a mailing list, contact the Mathematics Case Methods Project, WestEd, 300 Lakeside Dr., Oakland, CA 94612. To learn more about our project and other teacher professional development materials and activities visit the WestEd website at <www.WestEd.org>.

WRITING COLLABORATORS

Lori Albee

Susie Alldredge

Leanna Baker

Cynthia Bunker

Melinda Céspedes

Ruth F. Clark

Sandra Conley

Michelle Covarrubias, NBCT

Brandy De Alba

Holli Hall

Karla Heleotes

Kim Hertzog

Nancy Johnson

Tammera Johnson

Amy Kari

Melanie Maxwell

Paula Dyste Rothling

Introduction

What Do We Learn By Discussing Mathematics Cases?

Many professions, including business, medicine, and law, have used cases as learning tools. In education, talking about the dilemmas of teaching portrayed in a case is a rewarding way to learn from colleagues. It also helps us build analytical skills that, in turn, help us be more deliberate when planning and presenting lessons. One teacher who was a member of a case discussion group describes the process this way:

> Everybody in the group is looking at the same experience or piece of information, so what you get is everybody's input into what they see, what the situation means and, most importantly, how to change the situation to help a child learn.

According to the *Principles and Standards for School Mathematics* (National Council of Teachers of Mathematics 2000, 19): "Opportunities to reflect on and refine instructional practice are crucial." However, it is difficult to analyze our own teaching with detachment. For one thing, what we believe and assume about teaching and learning clouds much of what we do.

One benefit of using cases as learning tools is that it's easier to stand back and look at someone else's teaching in a more objective way; this can open the door to learning how to, and being willing to, analyze our *own* teaching with a critical eye. During discussions, we become aware that others may see a situation different from the way we do. A facilitator, whose job is to orchestrate the case discussion process, is essential in helping the group critically examine the individuals' beliefs and the ideas in the case. The group's responsibility is to dig into the ideas, weigh the trade-offs, and be open to changing their point of view. Sometimes, some or all group members end up changing their thinking after discussing a case. On the other hand, we may reconfirm our thinking and come away with a stronger rationale for what we already do and believe.

Each of the cases presented here is designed to promote analysis of four elements: (1) mathematics, (2) student thinking, (3) instruction, and

(4) language. This, again, is in keeping with one of the principles outlined in the *Principles and Standards for School Mathematics:* "Effective teaching requires knowing and understanding mathematics, students as learners, and pedagogical strategies" (National Council of Teachers of Mathematics 2000, 17). The unique characteristic of case discussions is that these elements are considered interactively, rather than independently, which is the way they occur in teaching practice. The knowledge gained from understanding the interplay among these elements, sometimes called *pedagogical content knowledge,* is at the heart of teaching.

More specifically, the cases in this book are intended to:

- Deepen our understanding of the mathematics we are teaching students.
- Help us experience the mathematics from students' points of view so that we understand what is confusing and why.
- Help us, respectfully and sincerely, examine various teaching strategies for their benefits and drawbacks.
- Help us understand how oral, symbolic, and written communication impacts the mathematics learning of all students.

By attending to these interconnected elements, we can better understand and evaluate how various manipulatives, tasks, questions, symbols, words, and drawings can facilitate learning or contribute to misunderstandings. In addition, the cases may help us develop multiple strategies to compensate for, and work through, any misunderstandings that occur. Ultimately, lessons need to incorporate a more cohesive and targeted plan for what we want students to learn, as well as thoughtful strategies for helping them to achieve goals and to assess their understanding.

What Cases Are and How They Are Used

As teachers, we make thousands of decisions daily. Few of these decisions are clear-cut; usually, they require making trade-offs. The cases in this book provide opportunities for teachers to talk about some of the important trade-offs involved in teaching beginning numbers and operations in the primary grades. Written by primary grade teachers, each of the case stories describes a classroom experience that had an unexpected outcome or was surprisingly difficult. The cases, each containing dialogue and student work, describe how the instruction was planned and what actually happened. Because they have been written by teachers who teach in a broad range of school communities, the cases offer opportunities for diverse groups of teachers to learn from and with each other.

Cases are used in many ways. Sometimes they are part of a more comprehensive professional development program. For example, because case discussion is especially effective in helping teachers develop the capacity for more in-depth analysis of student work, cases may be incorporated into a professional development effort aimed at learning to assess student work.

Cases that describe experiences with mathematics also motivate teachers to learn more about both the mathematics they teach and students' thinking about what's being taught. We find that these cases are particularly effective when teachers are also reading some of the research about student mathematical thinking. For this reason, we have included a Read and Reflect section at the end of each case. Teachers in discussion groups can gain more knowledge about the mathematics, student thinking, and pedagogy in the cases if they choose to use these additional resources.

The cases in this book can also be used as a basis for school-site discussion groups. Typically, a group of 6 to 12 teachers (often from a wide mix of elementary and middle school grade levels) meets once a month for about two hours to discuss a case. Members of discussion groups usually plan to discuss a collection of cases over time and often choose to revisit a case several times to gain a greater depth of understanding. A brief description of a process for discussing a case can be found later in this chapter.

What These Cases Are About

The mathematics topics embodied in these cases are central to the primary grade curriculum, regardless of the instructional materials used. They also correlate easily to standards and assessments of student learning whether they are at the national, state, or local level.

Mathematics Content and Alignment with Standards

A common difficulty is that content standards for students are too distant from the classroom to accomplish their purpose, which is to improve student performance. By linking discussions of the standards to case discussions, teachers can achieve greater clarity about what a standard means. Also, once the standard, or target for learning, is clearly identified and fleshed out, case discussions help us evaluate ways to help students achieve that standard. Finally, case discussions provide many instances of how students might misunderstand the ideas related to a standard so that those confusions can be worked through and overcome.

One way to give substance to a standard is to set aside reflection time to talk about what can be learned from cases and how that contributes to our understanding of that standard. For example, we might ask: "What did we discuss that helps us know if a student is performing well with regard to this standard?" and "What teaching strategies did we talk about that are important for this standard?" In addition, we might use what we learned from cases to look through our instructional materials and tasks and ask ourselves whether they are aligned with what we want students to learn, and if they require rigorous thinking on the part of students. In these ways, explicit reflection on relating the cases' standards to teaching and learning can ensure more purposeful instruction.

To facilitate alignment of case discussions with standards and student learning assessments, we provide a content matrix that highlights the specific mathematics topics within and across the cases (see p. 5).

Crosscutting Themes and Case-Specific Issues

The cases were also developed in ways that would point out six important crosscutting themes, which revolve around ideas and decisions that teachers need to talk through from various perspectives; unfortunately, they seldom have the opportunity to do so. The themes in the cases invite teachers to seriously examine, and possibly modify, commonly held beliefs and prevalent practices. The interplay between mathematics, student thinking, instruction, and language is also drawn out through these themes.

While there is overlap across the themes, there are also subtle differences. Each theme is described in the following sections.

Relationships Among, and Boundaries of, Mathematical Ideas   This theme draws into sharp focus the fact that students, and sometimes teachers, overgeneralize mathematical concepts or define them too narrowly or ambiguously. Similarly, students may notice relationships among mathematical ideas but may or may not have opportunities to fully articulate their meaning or understand their significance. For example, students may learn that *turnaround* addition facts, such as 4 + 3 and 3 + 4, are both equal to 7. Then they may incorrectly apply the same idea to subtraction facts and think that

Mathematics Content Matrix

Cases	Number Sense				Number and Place Value Concepts				Addition Concepts and Skills				Subtraction Concepts and Skills				Multiplication Concepts and Skills				Algebraic Thinking		
	Numerical flexibility	Relative magnitude	Reasonable answers	Rounding	Counting	Value of money	Base ten number system	Grouping and regrouping	Model/solve word problems	Model/solve symbolic problems	Basic facts	Multi-digit computation	Model/solve word problems	Model/solve symbolic problems	Basic facts	Multi-digit computation	Model/solve word problems	Model/solve symbolic problems	Multiples of ten	Multi-digit computation	Equality concepts	Properties	Equations and expressions
You Just Count the Extras	X		X		X				X	X			X	X									
Word Problems First, Then Basic Facts		X							X	X	X		X	X	X								
Everything in Its Place		X	X		X		X	X					X	X		X							
Trading Games							X	X				X				X			X				
Stop Sign or Equals Sign?	X																				X	X	X
Daily Equations	X									X	X			X					X		X	X	X
Diana	X				X					X													
Tallies and Coins	X	X			X	X	X																
How Many More?			X										X		X						X	X	X
Why Is Subtraction More Difficult?	X				X				X	X		X	X	X	X	X					X	X	X
Tina	X			X					X	X	X	X	X	X		X			X				
Is Number Sense Enough?			X	X			X	X											X				
Closer to 500 Than 400	X	X	X	X			X												X				
Dollars and Cents Confusions	X	X	X	X		X	X	X	X	X			X	X			X	X	X	X			
Carry Two or Twenty?	X	X	X	X	X		X					X				X	X	X	X	X			
How Many Can She Buy?	X			X	X	X			X			X	X			X	X	X	X	X			

4 − 3 and 3 − 4 are both equal to 1. This often occurs because the *commutative property*—the mathematical basis for *turnaround* facts—was not carefully discussed to help students define and apply it more explicitly.

Another example is when subtraction is narrowly defined as *take-away*, which leads to difficulty in understanding why you could subtract to find the *difference* in the weights of two puppies. After all, nothing is being taken away so how could it be associated with subtraction? This narrow concept of subtraction also may interfere with understanding the relationship between *compare* problems and subtraction, or subtraction and missing addend number sentences.

Extracting Mathematical Meaning from Words, Models, and Symbols
What do students *say* to themselves when they read symbols or word problems? How do they interpret the manipulative models we provide for them? How does this impact what they think and do? This theme helps us think through ways to unwrap and draw out children's internal chatter, and to understand the importance of attaching mathematical meaning to words, manipulative models, and symbols.

Likewise, we can discuss the benefits and limitations of various strategies for helping students understand word problems such as asking them to focus on key words, to restate the problem in their own words, or to make a diagram to represent the problem. The analysis of the interplay between verbal and symbolic mathematical language has the potential to impact the precision with which teachers approach mathematical talk in their classrooms, and the care with which they choose mathematical tasks and models to teach a concept.

Developing Flexibility and Efficiency with Numbers   This theme provides the setting for discussing the tensions that arise between helping students develop flexibility and efficiency with their mathematical thinking and computation. The trade-offs of various teaching strategies for developing flexibility or efficiency come into focus through this theme. For example, one case brings up the issue of whether students should be expected to use their number sense or a standard algorithm to solve a problem such as 97 × 4. Another case examines why students can do a problem with manipulatives, but don't transfer this knowledge to solving a problem without manipulatives.

Intentional Teaching and Learning   The crux of this theme has to do with clarifying the mathematical goal of a lesson for ourselves, making sure that the goal is understood and shared by students, and aligning tasks to the goal. A theme that runs through cases is that children often have the mistaken goal of learning a game, procedure, or pattern in lieu of the mathematical goal of the task. In other instances, we find that what the teacher intended to teach does not match the tasks that the students are doing.

Prompting Answers or Provoking Thought   The purpose of this theme is to prompt deep analyses of the mathematical questions, manipulative tasks, or problems that appear in the cases and to ask, "Does this task or question require students to think hard, or hardly think at all?" One case, for example, prompts teachers to analyze a place value task using base ten blocks to see if students can do the task by simply following a pattern rather than learning about place value concepts. Another case brings up debate about the benefits and limitations of constructing tasks that students are unlikely to get wrong, versus tasks that might invoke a pitfall in student thinking.

Equalizing Participation and Learning   This theme is a catalyst for asking how to alter instruction to ensure that all students participate and learn. The cases motivate us to explore multiple ways of representing mathematical concepts so that if one way doesn't work for all students, another way, perhaps using a different modality, can be tried.

We also examine how the language used to discuss or represent mathematical ideas impacts learning, especially for students less proficient in the English language. This theme calls attention to ways teachers might vary participation structures so that students are comfortable taking risks and talking about their thinking, correct or flawed. It also challenges teachers to think of ways to assist students who need extra support, without watering down a mathematical task or betraying its mathematical integrity.

In addition to laying out crosscutting themes, we have also identified sample specific issues that are related to the themes for each case (see pp. 9–12). The issues are the backbone on which the case discussion pivots. The issues prompt discussion about mathematical representations and concepts, student work, dialogue, and teaching decisions illustrated in the case. It is important to realize, for example, that a particular student's solution to a word problem is shown in the case for a reason, and that her work will help bring an important issue to light. In other words, each case selectively pulls a few things from the many possible examples of dialogue or student work. Therefore the discussion of a case is purposefully crafted to address key issues.

Some of the issues raise questions about the alignment of an activity or a task with a mathematical goal. Other issues prompt case discussion participants to weigh the benefits and limitations of a particular manipulative or task. Some issues invite debate about pedagogical decisions, while others stimulate discussion about a mathematical concept or pitfall related to that concept.

The issues are written in question format to emphasize that they are stimuli for discussion and debate. By writing the issues as questions rather than statements, we are more likely to elicit different points of view and alternatives. We are, hopefully, modeling how teachers might frame the issues for themselves as part of their own reflective practice.

Less experienced case discussion groups may find it helpful if the facilitator writes this book's sample issues on a chart in front of the group. Then, prior to beginning the discussion, teachers in the group could generate their own issues to add to the list. As teachers gain experience with the cases, and learn how to better identify and formulate key issues, they may prefer to generate all of the issues themselves, rather than using those from the casebook. Included in the following table is an overview of some key issues the cases address. For convenience, the issues are also listed in the facilitator's guide notes for each case.

Explicit Attention to the Role of Language

The cases in this book also focus on language as an important component of teachers' pedagogical content knowledge. This overt emphasis on language results in careful attention to what a teacher in a case says (or could have said) and how the choice of words affects student thinking. Language plays a crucial role in these cases because it is in the primary grades that children encounter mathematical terms and symbols for the first time, while they are simultaneously developing language fluency in other academic and social areas.

During case discussions, teachers grapple with primary-grade mathematics concepts on an adult level. In doing so, they begin to realize how their own mathematical understanding is intertwined with the language in a problem or task, the mathematical terms and symbols, and dialogue with others. This experience helps them understand how the language they use during instruction influences the meaning their students will construct through their own language and thoughts. This realization leads many teachers to become more mathematically precise and explicit in their explanations. It also helps them understand the need for students to have many opportunities to express mathematical ideas orally, symbolically, and in writing.

A focus on language is evident in the *Attaching Mathematical Meaning to Words, Symbols, and Models* crosscutting theme, which recurs throughout several of the cases in this book. In addition, the case issues that are closely tied to language are called out in the Facilitator's Guide Notes by the *Language Links* and *Language Confusion* headings. These headings alert the facilitator to the fact that the mathematical topics embedded in certain parts of the case have important ties to language.

It is our intent that, through close scrutiny and critical discussion of the impact of language on mathematical thought, teachers will become aware of pitfalls in their lessons and gain knowledge about ways to support the mathematical language development of their students.

Case Discussion Process

The Mathematics Case Methods Project has developed a process for using the cases based on more than a decade of experience. Each component of the process is guided by the goals of our project and has been continually revised over time. One goal is that case discussions should be an inclusive activity during which diverse groups of teachers—reform-oriented and more traditional, confident and not so confident, math enthusiasts and math phobic—learn with and from each other. Another goal is to have discussions that challenge thinking about mathematical ideas, not just instruction. A third goal is to have discussions that are genuine, sincere, and respectful so that everyone in the group feels their views are given serious consideration. It needs to be an environment in which case discussion participants are willing to put things on the table to be evaluated for both benefits and limitations, and then be open to changing their minds.

Others have also developed processes for discussing cases that would be equally effective to this one. How the discussion is designed could depend on the experience of the teachers, whether the discussions are site-based or part of a more formal professional development, or how accustomed the teachers are to discussing with a critical eye.

The process described here has been used primarily with experienced teachers and has been successful with teachers regardless of their professional development experience, their mathematical expertise, or their

Crosscutting Themes and Issues

Case Titles	Crosscutting Themes	Mathematics, Student Thinking, Instruction, and Language Issues
You Just Count the Extras	Prompting answers or provoking thought Extracting mathematical meaning from words, models, and symbols	How does the action or relationship represented in a problem dictate how a student might draw or think about it? Why might students add instead of subtract when solving "How many more" problems? How can the relationship between subtraction and problems that ask "How many more?" be made more apparent to students?
Word Problems First, Then Basic Facts	Intentional teaching and learning Extracting mathematical meaning from words, models, and symbols Developing flexibility and efficiency with numbers	How does solving word problems differ from solving problems written in symbols? What are the benefits and drawbacks of starting with word problems instead of basic facts? What is the role of understanding and practice in learning basic facts?
Everything in Its Place	Intentional teaching and learning Prompting answers or provoking thought Extracting mathematical meaning from words, models, and symbols	How could we find out if students are doing place value tasks without understanding the value of the digits or the whole quantity? Is the hundred square block 1 or 100 to students? Why might it be confusing to students to ask "How many tens are in 152?"
Trading Games	Intentional teaching and learning Prompting answers or provoking thought Extracting mathematical meaning from words, models, and symbols	How do the *ticket* and *Race for Zero* activities differ? What is the mathematical goal of the lesson and how well does it match the activity? How do you connect manipulatives to abstract symbols?
Stop Sign or Equals Sign?	Intentional teaching and learning Relationships among, and boundaries of, mathematical ideas Extracting mathematical meaning from words, models, and symbols	How might students' early work with number sentences prompt them to think of the equals sign as a stop sign? Is it correct to have more than one equals sign in an equation?

Case Titles	Crosscutting Themes	Mathematics, Student Thinking, Instruction, and Language Issues
Daily Equations	Intentional teaching and learning Relationships among, and boundaries of, mathematical ideas Extracting mathematical meaning from words, models, and symbols	How might students learn that you can turn around addition but not subtraction problems? What does the equals sign mean to a student who writes an equation like $9000 - 9000 = 0 + 8 = 8$? How might a teacher capitalize on students' intuitive number sense?
Diana	Intentional teaching and learning Extracting mathematical meaning from words, models, and symbols Prompting answers or provoking thought	How do we assess whether students' solutions are based on understanding or the superficial application of patterns? What role does language play in understanding addition number sentences? What are the benefits and limitations of always having each activity move from manipulatives to symbols? How is the thinking required of students different for each of the addition activities in the lesson?
Tallies and Coins	Intentional teaching and learning Prompting answers or provoking thought Developing flexibility and efficiency with numbers	Why do students easily recognize that 1 bundle and 3 straws are 13 straws, but have difficulty knowing the cost of two items, one 6 cents and the other 10 cents? How were the counting tasks in the lesson similar and how were they different? How might the place value and counting routines using the calendar and tallies help or hinder children's flexibility with numbers? What are the benefits and limitations of prearranging quantities from greatest to least for children to count or add?

Case Titles	Crosscutting Themes	Mathematics, Student Thinking, Instruction, and Language Issues
How Many More?	Extracting mathematical meaning from words, models, and symbols Prompting answers or provoking thought	How do students interpret the phrase "How many more"? What are the benefits and drawbacks of using *guiding questions* versus *leading questions*? Why might a student write an addition number sentence for a *how many more* problem, even though they solved it by subtracting?
Why Is Subtraction More Difficult?	Extracting mathematical meaning from words, models, and symbols Developing flexibility and efficiency with numbers	What are the benefits and drawbacks of timed basic fact tests? What is the role of understanding and practice in learning basic facts? Why is subtraction more difficult?
Tina	Extracting mathematical meaning from words, models, and symbols Prompting answers or provoking thought Developing flexibility and efficiency with numbers	Why might asking students to prematurely record their thinking undermine intuitive understanding? How does the action or relationship represented in a problem dictate how a student draws or thinks about it? What are the benefits and drawbacks of encouraging students to use informal methods for adding before more formal procedures?
Is Number Sense Enough?	Extracting mathematical meaning from words, models, and symbols Prompting answers or provoking thought Developing flexibility and efficiency with numbers	What are the benefits and drawbacks of simultaneously teaching addition and subtraction from the beginning? What are the benefits and limitations of always starting with the *ones* column when we do multi-digit addition or subtraction? When and how do you introduce a pencil-and-paper algorithm? How do you connect a student's number sense to an algorithm?

Case Titles	Crosscutting Themes	Mathematics, Student Thinking, Instruction, and Language Issues
Closer to 500 Than 400	Intentional teaching and learning Equalizing participation and learning Relationships among, and boundaries of, mathematical ideas	What does *closer to* mean to students? How do we assess whether students' solutions are based on understanding or on superficial pattern finding? What are the benefits and limitations of using Hundred Charts or Count-by-Tens Charts? How can you encourage both cooperation and correct mathematics?
Dollars and Cents Confusion	Intentional teaching and learning Equalizing participation and learning Extracting mathematical meaning from words, models, and symbols	What are the benefits and limitations of using money to work with multiples of ten? What are some alternate ways Benito could participate in lieu of the whole-class discussion? How do students translate the value of coins into numerical quantities, and what are the likely pitfalls?
Carry Two or Twenty?	Prompting answers or provoking thought Developing flexibility and efficiency with numbers Equalizing participation and learning	What are the benefits and limitations of asking students to invent their own strategies or algorithms? What role might *standard algorithms* play in math discussions? What are some ways that *quiet* students can be given opportunities to participate in discussions?
How Many Can She Buy?	Developing flexibility and efficiency with numbers Relationships among, and boundaries of, mathematical ideas Equalizing participation and learning	Why might the problem prompt students to use repeated addition instead of multiplication? How do multiplication and division work as shortcuts for repeated addition and repeated subtraction? What are the benefits and drawbacks of putting an incorrect solution on the board for discussion?

teaching philosophy. We allow about two hours to discuss one case; the participants are instructed to read the case before the meeting. We also like to allow additional time for reflective activities connected to the case such as reading and discussing a paper about mathematics or research on student thinking. The following sections describe and provide a brief rationale for six components of our discussion process.

Inclusion Activity

We begin with a quick inclusion activity to help members relax, know one another on a more personal level, and become more willing to share opinions and ideas in the group.

Starter Problem

Teachers work individually on a mathematical problem (called a Starter Problem) that is posted by the facilitator from the facilitator's guide notes. While teachers work on the Starter Problem, they are asked to pay attention to their thinking and to consider what might be difficult or confusing from a learner's perspective. Insights and strategies are shared later when the case discussion begins, not immediately. This allows teachers, who may have made a mistake in the Starter Problem, to bring up the error as insight into the case, where it can be discussed in context.

Listing Facts

Teachers quickly call out relevant factual information and the facilitator lists it on a chart. The purpose of this component is to get teachers started talking and to bring relevant details in the case to the foreground. For example, a *fact* might be the grade level of the students in the case.

Generating Issues

Teachers are asked to work in pairs to generate issues in question form. We specifically ask that the issues be about the mathematics, student thinking, instruction, and language, with a high priority on the math. The facilitator may choose to post a couple of issues from the guide notes as examples; however, having teachers learn how to generate significant questions themselves is a valuable experience. After about 10 minutes, the facilitator records the questions from each pair of teachers on a chart.

Discussion

To begin the discussion, the facilitator asks for a volunteer to choose an issue and talk about their ideas to the group. During the discussion participants illustrate their thinking by drawing or writing their ideas on a chart or by demonstrating with materials. This slows the discussion, giving everyone time to digest what is being discussed. Both the participants in the group and the facilitator are responsible for bringing up issues, but it is critical to go into a few issues in depth rather than discussing a lot of issues superficially.

Closing Experience

We have tried a variety of reflective and summarizing experiences to close the discussion, but teachers usually want to continue the discussion rather than conclude it. Reflective activities (oral and written), however, have been successful in wrapping up a session, depending on the group and the time of day. More recently, we have found that asking teachers to read the facilitator's guide notes for the case is an appropriate reflective experience. It helps them pull together the ideas from the discussion and provides a frame of reference for further thought.

Using the Casebook and the Facilitator's Guide Notes

This casebook for primary grade teachers joins its counterpart for upper elementary and middle school teachers—*Mathematics Teaching Cases: Fractions, Decimals, Ratios, and Percents* (Barnett 1994). Both books (with facilitator's guide notes) were developed under the leadership of the Mathematics Case Methods Project staff at WestEd in San Francisco, California.*

Casebook Audience

Anyone who is interested in the teaching and learning of mathematics is likely to get hooked on case discussions. A careful introduction to the case discussion process by someone already familiar with it—whether a trained facilitator or a teacher who has experienced the process—helps get a case discussion group off to a productive start. It also helps if the facilitator has studied specific discussion leadership techniques and understands the issues in the case very well. It commonly takes participating in three or more discussions for teachers to fully understand and appreciate the whole process. The mental stimulation and the chance to discuss both practical and theoretical teaching problems with colleagues are gratifying.

The particular cases here were created by primary grade teachers and will be most appropriate for teachers of those grades. However, many teachers from upper elementary and middle grades have told us that these cases are also stimulating for them. After all, many of the same issues apply whether, as teachers, we are trying to help students understand subtracting with whole numbers or subtracting with fractions or decimals.

---

* The Mathematics Case Methods Project also offers professional development and materials to help case discussion groups get started.

This casebook was conceived primarily as a learning tool for experienced teachers. However, some teacher-preparation programs have successfully used our rational number casebook with preservice teachers. We believe that one of the reasons preservice teachers can have successful discussions is that they can readily draw on their own experiences of *learning* mathematics to consider the mathematics and instruction from the students' point of view.

There are two ways to use this casebook. The cases are intended to stimulate collaborative reflection through discussion; however, they can also be read, along with the facilitator's guide notes, to stimulate individual thought. Reading the cases with neither discussion nor the facilitator's guide notes is less likely to be productive because you will be limited to your own perspective and ideas. It is the interaction with the ideas of others—ideally, through discussion, but also by reading the facilitator's guide—that evokes new perspectives and raises questions about the status quo.

The cases are designed to spark a strong appetite for teachers to more fully understand mathematics. However, in the beginning, teachers may dismiss the mathematics in the case because they can overlook the complexities of it when it is seen through a child's eyes. We find that, especially for the first few cases, teachers appreciate reading the facilitator's guide notes after a discussion. They report that it helps them put the mathematical understandings they are developing into words. The notes are carefully written to provide mathematical background on each of the topics highlighted in a case. Although this may seem like *giving away the answers,* teachers describe it as more of a reflective experience—one that promotes rather than limits further thinking.

Guide Notes at a Glance

Each case in this book is followed by facilitator's guide notes. For ease of use, each set of notes is laid out in the same way. The purpose and rationale for each section of the notes is outlined next.

Case Synopsis   Each set of notes begins with a synopsis of the case. This brief description helps a facilitator, case group leader, or teacher educator determine which case, or set of cases, best suits a particular audience at a particular time. It also sets the stage for reading the guide notes that follow.

Sample Discussion Issues   Each case has been carefully analyzed to identify three or four key issues. They are presented in this section of the facilitator's guide notes as well as in the Crosscutting Themes and Issues matrix for the complete set of cases (see pp. 9–12). One purpose of the Sample Discussion Issues is to help the facilitator anticipate and prepare for important discussion points. Also, prior to a case discussion, the facilitator may want to post a sample issue and then invite the group to generate additional issues. Ideally, over time, the discussion group will learn to identify key issues in the case as they gain experience and will not need to rely on sample issues provided by the facilitator.

Suggested Materials   This section provides a list of suggested materials that are easy to collect and are commonly used in classrooms. Teachers in case discussion groups can use the materials to illustrate a point, to demonstrate an alternative way of teaching a concept, or to help them work through a mathematical problem that comes up.

Starter Problem   The Starter Problem functions as a vehicle for teachers to get their heads into the mathematics of the case and also to facilitate looking at mathematics through a child's eyes. The Starter Problem is usually copied onto a chart and teachers work on it before reading or discussing the case. We have found that the use of a Starter Problem enriches the discussion by focusing the group on mathematics and student thinking during the initial stages of the case discussion process.

Discussion Preparation Notes

The *meat* of the facilitator's guide notes is provided in succinctly written paragraphs that describe the mathematical, pedagogical, student learning, and linguistic issues embedded in the case. Mathematical background is given a high priority; rather than being separated out, it is situated in the instruction and student learning portrayed in the case. This section of the notes helps the facilitator anticipate the topics and issues that are likely to come up in a discussion. More important, it helps the facilitator study, plan, and set goals for the discussion of a particular case by linking the issues and background with examples taken directly from the case.

1. You Just Count the Extras

Last fall, my second graders were practicing addition and subtraction fact families and working on problem-solving skills involving both addition and subtraction. While I was out of school for a few days, the substitute teacher, following our district-adopted textbook, taught students to subtract when the problem asked "How many more?" After returning to school, I decided to check their understanding of this problem type by presenting the following:

> Brandon had 14 lollipops. Jeffrey had only 6 lollipops.
> How many more lollipops did Brandon have?

The students quickly got to work, using a variety of manipulatives to solve the problem. I then asked them to record their individual solutions. While they worked, I walked around and talked with them about their strategies. To my amazement, most of my students had immediately added the numbers. Steven, who finished first, explained, "I put 14 in my head and counted up 6 more."

Tona agreed with Steven, "I got 20 too. I drew all the lollipops and added them." I was dismayed to find that most of the students agreed with them.

I was beginning to worry about so many students adding, then Brandon said, "I drew all the lollipops too, but I only counted 8. I know the answer is 8." His page just showed a row of 6 lollipops at the top and 14 more scattered below.

Denisha said she got 8 too, "I made the numbers with color tiles, and I put them beside each other, and I didn't count the ones that match. I counted 8."

Alex, looking at the tiles on Denisha's desk, insisted, "You have to count all of them, because you have more than 8." Evidently Alex assumed that when you have two rows of cubes you count them all like you do when you

© 2003 by WestEd from *Number Sense and Operations in the Primary Grades*. Portsmouth, NH: Heinemann.

Brandon had 14 lollipops. Jeffrey had only 6 lollipops. How many more lollipops did Brandon have? __*2 0*__

add. At this point, most of the class began to discuss whether to count all the lollipops or just some of them. Then Hector shouted excitedly, "I know how to show it! I made it with snap cubes, and I know it's 8 because you put Brandon's and Jeffrey's side by side and you just count the extras!" Hector held up his snap-cube graph for the class to see.

© 2003 by WestEd from *Number Sense and Operations in the Primary Grades*. Portsmouth, NH: Heinemann.

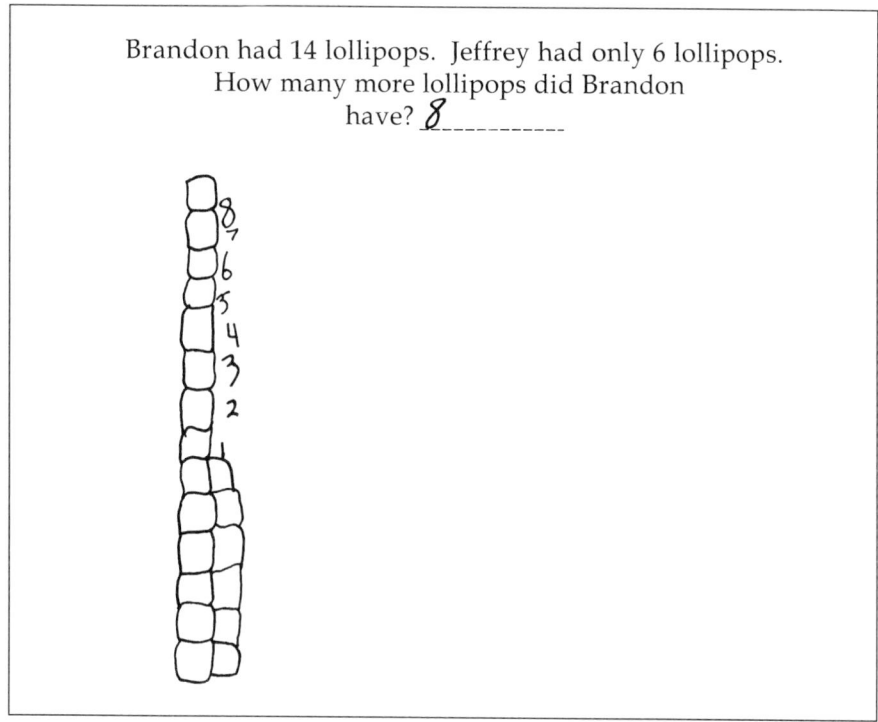

Immediate comprehension swept the room and several students re-peated, "You just count the extras."

Noticing that James had written 14 − 6 = 8 on his paper, I asked if he had solved the problem by subtracting. He replied that he drew the lollipops and counted them. When I asked why he had written the subtraction num-ber sentence on his paper, he quickly replied, "I drew 14 and covered up the 6 and had 8 more, and that's the same as 14 take away 6 equals 8." But look-ing at his paper, I couldn't tell if that was what he really did.

Still trying to help students make a connection with the previous text-book subtraction lesson, I asked if anyone else thought subtraction could solve the lollipop problem. No one replied. When I held up the textbook pages with the problems they had successfully completed the week before by using subtraction, Heath said, "The teacher said we were supposed to subtract on those problems."

© 2003 by WestEd from *Number Sense and Operations in the Primary Grades*. Portsmouth, NH: Heinemann.

Brandon had 14 lollipops. Jeffrey had only 6 lollipops. How many more lollipops did Brandon have?_____

I counted from 6 to 14

Heath's explanation drew quick agreement from the other students, so I concluded the lesson by asking, "How do you solve a problem that asks 'How many more?'" Most of the students immediately replied, "You just count the extra ones."

I can't help but wonder why so many of my students automatically added the numbers, even though they had subtracted the previous week. When Hector had explained his snap-cube graph, most of the class seemed to understand how to get the answer, but I don't think they really saw what he did as subtraction. I wanted them to realize that there is more than one type of subtraction problem so that they have something to fall back on when they get to compare problems with larger numbers, and counting the extras would not be practical.

Read and Reflect

Bay Area Mathematics Task Force. 1999. *A Mathematics Source Book for Elementary and Middle School Teachers: Key Concepts, Teaching Tips, and Learning Pitfalls.* Novato, CA: Arena Press.

© 2003 by WestEd from *Number Sense and Operations in the Primary Grades*. Portsmouth, NH: Heinemann.

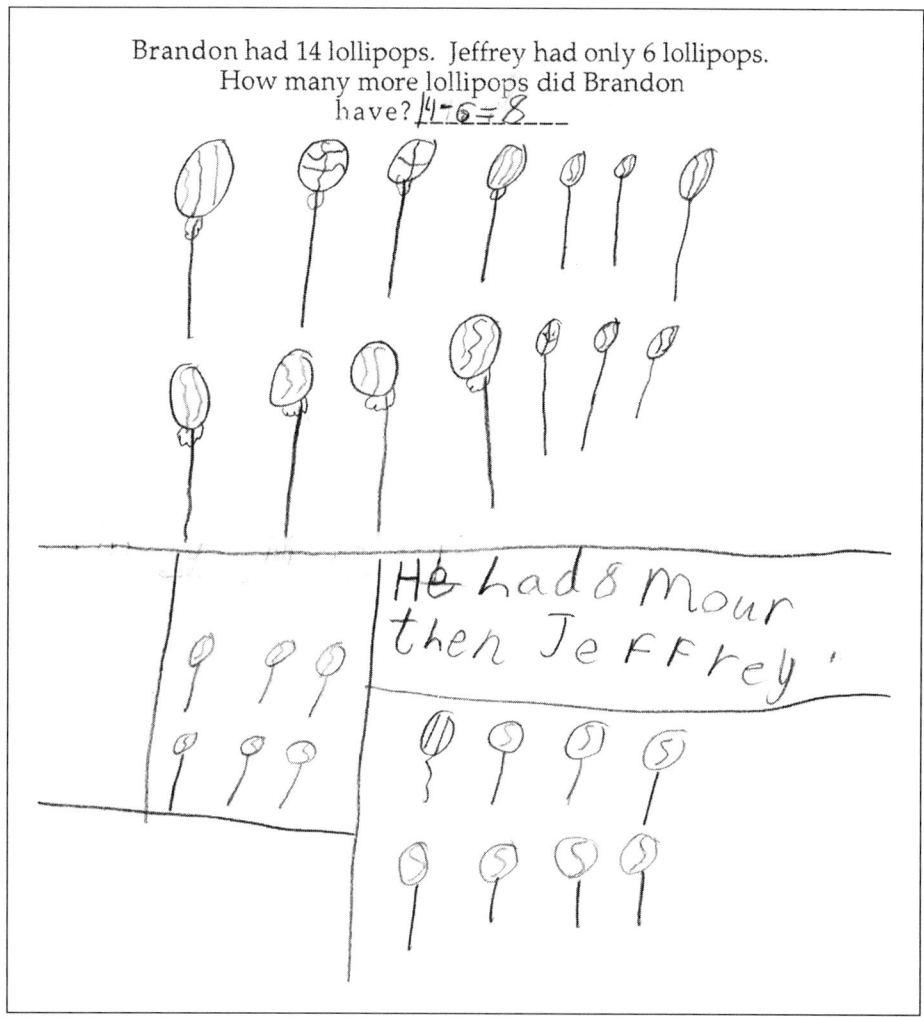

Brandon had 14 lollipops. Jeffrey had only 6 lollipops. How many more lollipops did Brandon have? 14-6=8

He had 8 mour then Jeffrey.

Read pages 14 through 17 on addition and subtraction concepts.

Trafton, P. R., and C. Midgett. 2001. "Learning Through Problems: A Powerful Approach to Teaching Mathematics," *Teaching Children Mathematics*, 7 (9). Read pages 532 through 536.

© 2003 by WestEd from *Number Sense and Operations in the Primary Grades*. Portsmouth, NH: Heinemann.

You Just Count the Extras
Facilitator's Guide Notes

A second-grade teacher gives students a comparison word problem to solve using any method that they choose. The students use a variety of materials to solve the problem and then explain their solution using drawings and writing. Most students, incorrectly, find the total of the numbers rather than the difference. The teacher doesn't understand why, especially since the students had successfully solved similar textbook problems the week before by subtracting.

Sample Discussion Issues

How does the action or relationship represented in a problem dictate how a student might draw or think about it?

Why might students add instead of subtract when solving *how many more* problems?

How can the relationship between subtraction and problems that ask "How many more?" be made more apparent to students?

Suggested Materials

Counters, color tiles, or linking cubes

Starter Problem

Draw a picture to show how you would solve this problem. Then write a number sentence.

Brandon had 14 lollipops. Jeffrey had only 6 lollipops.
How many more lollipops did Brandon have?

Language Confusion: *More* Means *Bigger*

This is a comparison problem, asking for the difference between two quantities. So why do most students in this case try to solve the problem by adding? Some students, like Steven, may associate *more* with getting bigger. Confusion with the word *more* may be compounded for Spanish-speaking students who may translate *more* as *más*, which can also be interpreted as *plus*.

In prior grades, students may have been given *keyword* strategies to help them solve word problems. For example, they may associate the keyword *more* with *bigger* or the addition operation. What are some ways the teacher can help students visualize the action or mathematical relationship portrayed by a problem? Is the phrase, "You just count the extras," a similar keyword strategy or a helpful way to think about the problem?

Subtraction Concepts: Modeling Comparison Situations

In school, students are often introduced to only one way of thinking about subtraction, the *take-away* model. Comparison problems, which require finding the difference, are also subtraction situations. Notice that James writes $14 - 6 = 8$ on his paper and explains, "14 take away 6 equals 8." However, James' drawing doesn't portray either the *take-away* situation he describes or the *comparison* situation in the problem. It appears that he may have written the number sentence first and then just drawn lollipops to represent each of the numbers.

There are big differences between modeling a *take-away* situation and a *comparison* situation. In a *take-away* situation, you start by counting out only *one* set. Then, the second set is taken away from the original set.

In a comparison situation, however, you usually begin by counting out *both* sets. Notice that addition situations are also commonly modeled by counting out two sets. This may trigger some students to revert to the more familiar action of *counting all* instead of finding the difference.

14 *take away* 6 Draw 14; then cross off 6.	o o o o o o o o ⌀⌀⌀⌀⌀⌀ 8 left
14 *compared* to 6 Draw 14 and 6. Match, and then count the extras.	ꝑꝑꝑꝑꝑꝑ o o o o o o o o ꝑꝑꝑꝑꝑꝑ 8 extras

Subtraction Concepts: More than *Take-Away*

In the latter part of the lesson, students seem to understand that you *count the extras,* but they still don't associate finding a difference with subtraction. It is one thing to get a solution and another thing to relate the subtraction operation to a given situation and solution process. Once students can solve a problem intuitively using common sense or materials, it is helpful to ask them to figure out which operation they can use to get the same answer. By solving many similar problems, they can be helped to make the generalization that one can solve comparison word problems such as these by subtracting.

Many students are exposed to subtraction as solely a *take-away* situation. By defining a concept like subtraction too simply, teachers run the risk of limiting the variation in the problems and tasks that students are able to do. As a result, when students are exposed to comparison situations with larger numbers, they might get *stuck* trying to use cubes or count up to find the difference. Students who have a broader concept of subtraction are more likely to recognize that a comparison situation can be solved by subtracting and have more flexibility in their thinking about different subtraction situations.

Notes

2. Word Problems First, Then Basic Facts

I have a second-grade class composed of mostly ESL students. It is now February, and I am starting to worry because so many of my students are still not on solid ground with basic addition and subtraction facts.

In first grade, they spent the year using a widely recognized manipulative-based curriculum that focuses on math facts up to 10. This year, we have spent the last three months practicing basic facts to 20 using daily word problems. We do simple *Problem of the Day* activities, in which the students solve problems, and then share their solutions and strategies with the class. After we solve the problems, students write in their journals about how they solved them. This is an example of a problem from Jennifer's journal.

My sister had 12 gumballs

She chewed 3. How many did she have left? 9

I counted backwards on my fingers,

After several students share their solutions, we talk about what number sentences we could write for the problem. I write the number sentences on the board, and the students copy them in their journals.

© 2003 by WestEd from *Number Sense and Operations in the Primary Grades*. Portsmouth, NH: Heinemann.

$$12 - 3 = 9 \qquad \begin{array}{r} 12 \\ -\ 3 \\ \hline 9 \end{array}$$

I want to be sure to connect the number sentences to language so I always read the number sentence to them. For example, I might say, "You could say 12 gumballs take away 3 gumballs leaves 9 gumballs. This is how you write number sentences when you are taking away."

Some other examples of problems we have done in class follow. I always give the students the problem, let them invent a strategy to solve it, and then review the addition or subtraction number sentences with them. Some of the tasks can be done by using manipulatives, and I often ask students to work on them with a partner. Students are able to choose their own manipulatives to use if they want; many choose linking cubes. But others dislike using manipulatives, preferring to draw. Still others decide to use their fingers or count in their head.

This group of partners decided not to use manipulatives or to draw anything. They just *did it*. When I asked Carlos and John how they had come up with their answer, Carlos said, "It's 14 days, because 7 days plus 7 days is 14. We just knew the answer."

I went to camp for two weeks.
How many days is that? 14 days.

Ana and Peter decided to illustrate their problem by drawing squares to represent chairs. I asked them what number sentence we should write and Peter said, "8 plus 20 is 12." I asked Ana if she agreed, and she did. Then I asked if it could be a subtraction problem. Both of them agreed that it wasn't a subtraction problem because there were 8 chairs and 20 chairs in their picture. I let it go since they had the correct answer and seemed to understand the word problem.

© 2003 by WestEd from *Number Sense and Operations in the Primary Grades*. Portsmouth, NH: Heinemann.

I have 8 chairs. There are 20 people in my class. How many more chairs do I need to give one to each person in the class?

$\Box\ \Box\ \ \Box\Box\Box\Box\Box\Box$ $\qquad$ $\Box\Box\Box\ \Box\Box\Box\Box\ \Box$
$\qquad\qquad\qquad\qquad\ \ \Box\ \Box\Box\Box$

12 more chairs

All of my students seem to enjoy these tasks, and, for the most part, they seem to understand the processes they were using. It appeared to be going so well that I decided to continue having them add to 20 and subtract from 20 by solving word problems instead of using the textbook to teach the procedures. Then recently, when I had a substitute, I left a worksheet of addition and subtraction facts for the students to do to make it easy for her. The substitute left the completed worksheets for me to see.

ADDITION AND SUBTRACTION

Room_____ $\qquad$ Name *Judy*

9	13	12	4	16
+ 2	− 5	+ 3	+ 7	− 8
11	12	15	11	12

10	5	8	3	15
− 6	+ 6	− 3	+ 8	+ 4
		11	11	19

It hadn't occurred to me that my students might have difficulty with the worksheet, but, in fact, few of them had been able to get the correct answers. I thought I had given them plenty of practice connecting number sentences to word problems, so I was surprised that they hadn't transferred their understanding of addition and subtraction to these basic computation facts. At this point, I don't plan to throw out the whole idea of beginning with word

© 2003 by WestEd from *Number Sense and Operations in the Primary Grades.* Portsmouth, NH: Heinemann.

problems because I see so much value in it. I just need a better way to help students transfer their understanding to worksheets or tests that don't involve word problems.

Read and Reflect

Bay Area Mathematics Task Force. 1999. *A Mathematics Source Book for Elementary and Middle School Teachers: Key Concepts, Teaching Tips, and Learning Pitfalls.* Novato, CA: Arena Press.

Read pages 18 through 22 on addition and subtraction concepts.

Geary, David. 1998. "Developing Arithmetical Skills," *Children's Mathematical Development.* Washington, DC: American Psychological Association.

Read pages 37 through 93.

Isaacs, Andrew C., and William M. Carroll. 1999. "Strategies for Basic-Facts Instruction." *Teaching Children Mathematics* 5 (9).

Read pages 508 through 515.

© 2003 by WestEd from *Number Sense and Operations in the Primary Grades.* Portsmouth, NH: Heinemann.

Word Problems First, Then Basic Facts
Facilitator's Guide Notes

The teacher in this case introduces addition and subtraction facts to 20 by presenting a variety of *Problems of the Day*. The students are expected to solve the word problems using any method or materials they choose, including manipulatives, drawings, or counting fingers. By and large, the students do well on these daily word problems, including writing out the number sentences that represent the solutions. But the students do not do well on a subsequent basic facts worksheet. The teacher wonders why they haven't transferred what they learned from solving word problems to this basic facts exercise.

Sample Discussion Issues

How does solving word problems differ from solving problems written in symbols?

What are the benefits and drawbacks of starting with word problems instead of basic facts?

What is the role of understanding and practice in learning basic facts?

Suggested Materials

Base ten blocks, play money, linking cubes

Starter Problem

Draw a picture to solve this problem, and then write a number sentence. Think about how a child would know what number sentence to write.

I have 6 pencils. There are 21 people sitting in my class. How many more pencils do I need to give one to each person in the class?

Language Links: Word Problems and Basic Facts

In this case, the basic facts have been taught in the context of solving word problems. Students solve the problems together, talk and write about solutions, and write the number sentences. They have been successful with these experiences, but their understanding has not transferred to basic fact exercises recorded as mathematical symbols. Why not?

Although students have translated problem solutions into symbols, they haven't worked *backwards*, imposing language and meaning onto symbols. Notice that solution methods can be naturally derived from interpreting word problems, but they are not so readily apparent in symbolic problems. Given a subtraction problem presented in symbols, students who don't know the answer right away must decide on an efficient method for finding the solution. That may involve comparing to find the difference, counting up from one number to the other, or putting up some fingers and taking some away. A *subtraction sign* could mean any of these.

Why Start with Word Problems?

Is there an advantage to having students start by solving word problems instead of solving symbolic problems with manipulatives first? One possible reason for beginning with word problems is that students come to school with intuitive mathematical knowledge and real-world experiences that they can draw on to interpret problems. Notice that students seem to get correct answers on the word problems but struggle with the symbols. Peter, for example, could solve the problem with the chairs, but thought that the number sentence would be $8 + 20 = 12$. Even after the teacher asked if it might be a subtraction number sentence, he and his partner were convinced that it was addition. Possibly the word *more* led them to think about addition instead of subtraction. How might the teacher help them connect this kind of word problem to subtraction?

Judy had incorrect answers for $13 - 5$ and $16 - 8$. For example, she got 12 as the answer for $13 - 5$. Perhaps she thinks of 13 as the individual digits, 1 and 3; since she couldn't take 5 from 3, she made it work by taking 3 from 5. She made a similar error for $16 - 8$. Yet if given a word problem that involved the same numbers, it is unlikely that she would have made the same mistakes.

How do we help children learn to solve symbolic problems in ways that provide links to their intuitive understandings? The teacher's method of writing the number sentence for the word problems is one way. However, children also need experiences writing their own number sentences for the problems they solve, as well as translating symbols into oral sentences and into word problems.

Language Links: Reading Mathematical Symbols

Students benefit from explicit discussion about the different ways symbols can be interpreted in words. For example, it may be helpful to have them practice reading "20 − 8 = ____" as "20 is how many more than 8?" and as "20 take away 8 is how many?" Also, notice that it is legitimate to interpret the chair problem in the case as a *missing addend* problem, which would be written symbolically as "8 + ____ = 20," and interpreted as "8 and how many more is 20." Younger students may need considerable guidance to develop these skills over time.

It may be tempting to avoid language-intense experiences, such as solving word problems or reading symbolic number sentences orally, with students who are poor readers, have weak language skills, or who are learning English as a second language. However, if language-rich experiences are omitted, students miss opportunities to develop meaning for symbols and connect mathematics to real-life situations. It is helpful to provide support by reading word problems to students, discussing important vocabulary, and having students share different ways of representing their solutions.

Role of Intention: Learning Basic Facts

Most students will eventually be expected to *learn* their basic facts. Will this occur from solving many word problems? Memorizing or establishing quick recall of basic facts is likely to be easier for students who have a firm understanding of the underlying concepts. However, this understanding does not necessarily translate into quick recall of basic facts. Most students also will need to practice with the explicit intention of mastering quick recall. The degree of repetition needed to develop a facility with basic facts will vary from student to student and from fact to fact. Students will benefit from experiences that focus on both committing facts to memory, through games or other practice, and learning how and when to use counting strategies to compute the answers.

3. Everything in Its Place

I teach at an urban school with a total enrollment of about 500. My class is a combination first/second grade with 30 Spanish/English bilingual students. Since the beginning of this year, we have been exploring the concept of place value in various ways.

One way has been by doing a daily calendar activity. Each day I write the previous day's date on the board, and, together, we add *1* to represent one more day. When the date changes from the 19th to the 20th or the 29th to the 30th, we talk about how the numeral in the tens place changes. This helps the children begin to understand the meaning of each place in a two- or three-digit number.

Place value has also been a part of our math exploration activities. One game we frequently play is called *Race for the Raft,* which is played with ones, tens, and hundreds blocks (small cubes, *rods,* and *flats*).

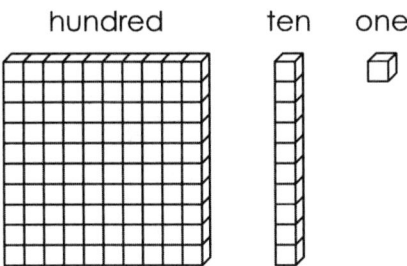

Before I introduce the game, I give the children a chance to handle and explore the blocks. After they've become familiar with them, I tell them that we call the small cubes *ones.* Then I ask them to see if they can figure out what the rods are called and why. After they have identified the rods as *tens,* we move on to the flats. I ask them to use the *ones* and *tens* to see if they can figure out what the flats are called. If necessary, we count the squares on the topside of the flat to establish that it represents *one hundred.*

The children play *Race for the Raft* working in groups of two or more. Each has a regular die; a board marked with hundreds, tens, and ones; and a

© 2003 by WestEd from *Number Sense and Operations in the Primary Grades.* Portsmouth, NH: Heinemann.

bucket of blocks. To start the game, each child takes a turn rolling a die, and places the corresponding number of cubes into the *ones* column. When a child has 10 *ones,* the ones are exchanged for a rod (or *ten*) and the child places the rod in the *tens* section of the board. As soon as a child has 10 rods, the rods are exchanged for a flat. The first player to get a flat (a *raft*) wins the game.

When they first start playing this game, I don't ask students to write down the numerals that correspond to the blocks they see on their board. After lots of oral practice, I begin to show them how to write the correspond- ing numbers. I begin by showing them how to write the numeral for a sam- ple game board like this one. First I ask, "What number did we make with the blocks?"

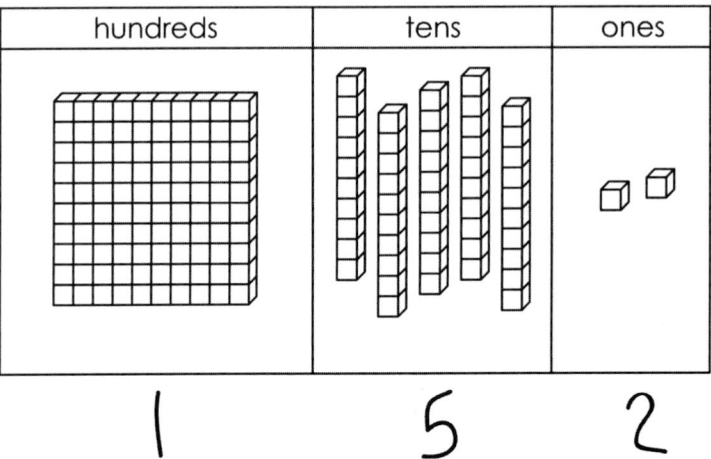

hundreds	tens	ones

1   5   2

"One hundred fifty-two," they answer, and I write the number on the board.

Then I ask questions like, "How many hundreds in 152?"

"One," they respond in unison.

"How many tens in 152?"

"Five," they answer back.

Then I ask, "How much is 5 tens?"

"Fifty," they call out.

"How many ones in 152?"

"Two."

"Good. Let's say the number all together." "One hundred fifty-two!"

© 2003 by WestEd from *Number Sense and Operations in the Primary Grades.* Portsmouth, NH: Heinemann.

From this game, it is an easy jump to writing numbers in expanded notation. Using another sample game board, I show them how to write the number represented on the game board two different ways—in expanded notation and as an entire number.

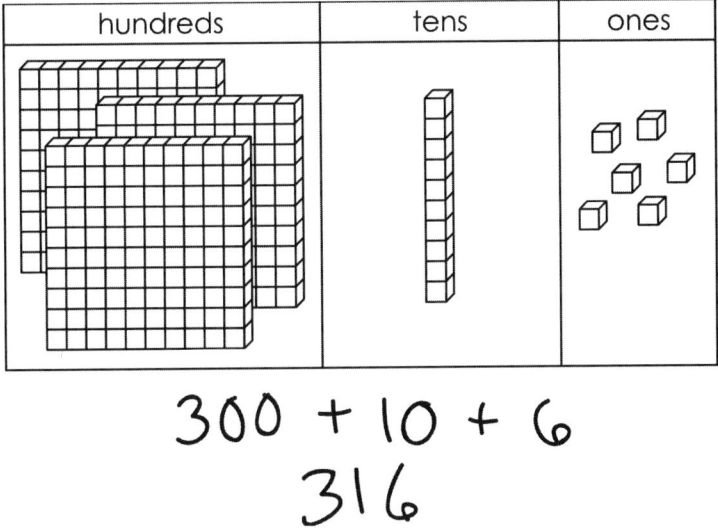

$$300 + 10 + 6$$
$$316$$

Based on their success with *Race for the Raft* and the lessons on writing numbers in different forms, I then gave them some practice problems to do on their own. Marianne's work on these problems caught me by surprise. But then I noticed that many other students made similar errors.

Marianne

hundreds	tens	ones
2	4	3

$$\underline{2} + \underline{4} + \underline{3} = \underline{9}$$

When I asked Marianne how she got her answer, she told me, "I just added 2 plus 4 plus 3 and got 9."

I wonder what they really learned from these experiences. It's almost as if they can *do* place value activities without having to think. They just look for a pattern. I wonder what kinds of activities I can use that won't be so predictable or obvious so that I can see what my students really understand.

© 2003 by WestEd from *Number Sense and Operations in the Primary Grades.* Portsmouth, NH: Heinemann.

Read and Reflect

Bay Area Mathematics Task Force. 1999. *A Mathematics Source Book for Elementary and Middle School Teachers: Key Concepts, Teaching Tips, and Learning Pitfalls.* Novato, CA: Arena Press.

Read pages 7 through 11 on number and place value concepts.

Cotter, Joan. 2000. "Using Language to Teach." *Teaching Children Mathematics* 7 (2).

Read pages 108 through 114.

© 2003 by WestEd from *Number Sense and Operations in the Primary Grades*. Portsmouth, NH: Heinemann.

Everything in Its Place
Facilitator's Guide Notes

The teacher in this case faces a dilemma. Although her students can do place value activities, they don't seem to apply the concept of place value to help them understand multi-digit numbers. The case provides opportunities to analyze the kinds of place value tasks commonly given to students, such as *Race for the Raft,* and to consider how they enhance or inhibit student understanding.

Sample Discussion Issues

How could we find out if students are doing place value tasks without understanding the value of the digits or the whole quantity?

Is the hundred square block *1* or *100* to students?

Why might it be confusing to students to ask "How many tens are in 152?"

Suggested Materials

Base ten blocks, play money bills, linking cubes

Starter Problem

Which of these expressions is the same as 312?

$$300 + 10 + 2$$

3 tens 1 hundred and 2 ones

$$10 + 300 + 2$$

$$3 + 1 + 2$$

Place Value Concepts: Individual Digits Versus Whole Quantity

In *Race for the Raft*, students practice trading hundreds for tens or tens for ones and explaining how many hundreds, tens, and ones they see on their place value board. Possible purposes for these activities are to help students understand the value of individual digits and to provide a foundation for regrouping in the future. Yet when a teacher places great emphasis on the number of blocks in each place, students may become married to the misconception that the digits in 324 are three separate numbers. The value of the *entire quantity* may be lost. Notice that instead of expanding 243 into 200 + 40 + 3 = 243, they simply write and add the individual digits.

Without first focusing on whole quantities, students may subsequently encounter problems regrouping when computing. For example, to find the sum of 34 and 26, they may add the digits 3 and 2 and the digits 4 and 6, coming up with two separate answers, a 5 and a 10, which they put together to get 510 as their final answer! A greater concern is that they may think this answer makes sense. How can the activities in this case (calendar activity, *Race for the Raft,* and expanded notation) be altered to make sure students don't lose sight of the whole quantity?

Language Confusion: 5 Tens in 152 or 5 in the Tens Place?

The teacher asked, "How many tens are in 152?" She seemed satisfied when students responded that there are 5 tens in 152. However, are 5 tens or 15 tens in 152? In this case, most likely the teacher meant to ask students, "How many tens are in the tens place?" The correct answer to this question would be "5 tens." Later, when students are studying division concepts, they might be asked, "How many tens are in 152?" The correct answer to this question is that there are 15 tens and part of another ten in 152. It is incorrect to say that there are 5 tens in 152.

Making Tasks Less Predictable and More Challenging

A key idea of place value is that the placement of a digit in a number makes a difference in its value. So a digit in the hundreds place is 10 times more than if it were in the tens place, 100 times more than if it were in the ones place, and so on. In activities such as *Race for the Raft* and the expanded notation board game, it is easy to assume that students understand place value because they can do exchanges—10 ones for 1 ten—or because they can write the correct numbers in the blanks for expanded notation. However, students may be successful in these tasks and have only a surface understanding of place value.

Race for the Raft is based on very predictable patterns. When students see 1 block, they write *1;* when they see 5 blocks, they write *5.* It doesn't matter what the blocks look like or what value they represent. Likewise, in the expanded notation game, students follow an easy pattern by writing a 3-digit number (316) using the first digit from each of the numbers in the expression (300 + 10 + 6). Once students understand the pattern, an incorrect response is almost impossible—as long as the pattern is fresh in their minds.

Might the task provoke more rigorous thinking if students were shown the blocks out of order (5 tens, 1 hundred, and 2 ones)? What if they were asked to prove that 50 + 100 + 2 is the same as 152? Perhaps students would begin to understand when order matters and when it doesn't. For example, while the 5 in 152 has a particular value because it is in the second place, the blocks can be presented in any order and their values will not change. Likewise, the numbers can be in any order in expanded notation and the value of the total will be the same. What kind of thinking would be provoked if students were given these problems?

$$327 = 20 + \underline{\quad} + 7$$

$$500 + 8 = \underline{\quad}$$

Notes

4. Trading Games

My class of 19 second graders spent seven weeks working with the lessons in a supplementary place value unit. After this, I taught them how to add one-digit numbers to two-digit numbers and then two-digit numbers to other two-digit numbers. At the end of four weeks, all but one of my students was proficient at using regrouping processes to solve and explain addition problems. They had demonstrated their understanding in a variety of ways, both with and without using manipulatives.

Once I felt comfortable with their abilities to solve addition problems with regrouping, I shifted into subtraction. The first two weeks were an overview of subtraction without regrouping. We did problems using manipulatives, then moved on to recording what we had done with the manipulatives. Finally, we moved on to worksheet practice. All students did well through this transition. The next logical step for instruction was subtraction with regrouping.

I began my regrouping unit with a whole-class lesson. Using the overhead projector, I led students through ticket trading, using base ten materials and a place value mat. A variety of prizes were displayed, and we assigned each prize a value of between 7 and 32 tickets. The class started with 75 tickets represented with base ten blocks. The object was to find how many tickets they had left after they *bought* each item. Students had had previous experience using these manipulatives with both addition and subtraction.

"You have 75 tickets to spend. Who can show me what 75 looks like?" Several students volunteered by raising *a quiet hand*. Angel showed the following on her place value mat.

© 2003 by WestEd from *Number Sense and Operations in the Primary Grades*. Portsmouth, NH: Heinemann.

Hundreds	Tens	Ones

"What would we like to buy first?" A student suggested the toy fish with a price tag of 32 tickets.

I asked, "Do we have enough tickets to buy the toy fish? Can we take 32 away from 75?" Students showed their agreement by using a thumbs-up signal. Lisa volunteered to show how many tickets would be left, using the materials on the overhead place value mat.

Hundreds	Tens	Ones

After buying two items, I realized I had not given them any problems with regrouping. Since we had 31 tickets left, I asked, "How about buying the hat? It's 25 tickets. Do we have enough tickets to get the hat?"

Students responded, "Yes, 31 is bigger than 25."

© 2003 by WestEd from *Number Sense and Operations in the Primary Grades.* Portsmouth, NH: Heinemann.

Brandon volunteered to show how to take 25 away from 31. But after studying the overhead, he finally said, "I can't. There aren't enough ones."

Hundreds	Tens	Ones

"Can you think of a way to get more ones?"

Jamie, one of the few who raised their hands at this point, said, "You can trade in a ten for ones."

"How many ones can I trade it in for?"

Several children said, "Ten."

Brandon then traded a ten for 10 ones, which he put in the *ones* column on the overhead place value mat.

Hundreds	Tens	Ones

© 2003 by WestEd from *Number Sense and Operations in the Primary Grades*. Portsmouth, NH: Heinemann.

"Do you have enough ones now?" He nodded and took away 2 tens and 5 ones, leaving 6 ones on the board.

Hundreds	Tens	Ones

After going over several similar examples as a class, students returned to their seats and worked with their own manipulatives to practice buying items. After questioning several students, and observing others as I walked around the room, I felt certain that everyone understood how to trade a ten for 10 ones when they didn't have enough ones. Numeric representation was not included at this point.

The next day, I introduced the *Race for Zero* game to give them practice trading and subtracting without recording. This game also uses base ten blocks and a place value mat. Starting with a flat of 100 on the mat, students rolled a die and then took that amount away from the 100. They kept rolling, each time subtracting the number rolled from the amount left on the mat until they reached zero. They played this game for a couple of days until I felt that all of the children were comfortable with trading.

Feeling confident about the children's understanding at this point, I decided it was time to connect numeric representation to what they were doing with the base ten materials. I chose to have them play the same game, but this time I asked them to write down numbers to show what they were doing. Their recordings were a mess! They had no desire to write the numbers down. They just wanted to play the game.

When I worked with individual students, they were able to record the number on the mat and the number on the die and, then write how many

© 2003 by WestEd from *Number Sense and Operations in the Primary Grades*. Portsmouth, NH: Heinemann.

they had left. Even then, though, their recordings were not well organized and didn't reflect the use of regrouping. The game was also hard to record because with each roll of the die, you take away a little bit. Some kids tried to record it as just one long problem. And many students wouldn't continue recording unless I was right there. I was so frustrated. I guess they either didn't understand it or just didn't want to do it.

I decided they needed more structure. At this point, I ditched the game and switched to prewritten problems that they were to solve using their base ten manipulatives and place value mats. I had students explain and show how the trading on the mats matched the *crossing out* in the numeric representation. The predictable six students quickly caught on, but the others were not there. I continued this method for a couple more days, but realized that no one else was catching on.

Finally, I turned desperately to a colleague to seek her ideas and feedback on this struggle to connect hands-on activities to math concepts and written work. I felt that I needed to reevaluate what I wanted my students to learn and how these activities did or did not fit that goal.

Read and Reflect

Bay Area Mathematics Task Force. 1999. *A Mathematics Source Book for Elementary and Middle School Teachers: Key Concepts, Teaching Tips, and Learning Pitfalls.* Novato, CA: Arena Press.

Read pages 12 and 13 on place value concepts and multi-digit computation.

Clements, Douglas H. 1997. "(Mis?)Constructing Constructivism." *Teaching Children Mathematics* 4 (4).

Read pages 198 through 200.

Fuson, Karen, Yolanda De La Cruz, Stephen T. Smith, Ana Maria Lo Cicero, Kristin Hudson, Pilar Ron, and Rebecca Steeby. 2000. "Blending the Best of the Twentieth Century to Achieve a Mathematics Equity Pedagogy in the Twenty-First Century." In *Learning Mathematics for a New Century.*

Read pages 197 through 212.

© 2003 by WestEd from *Number Sense and Operations in the Primary Grades.* Portsmouth, NH: Heinemann.

Trading Games
Facilitator's Guide Notes

A second-grade class spends many weeks engaged in carefully selected activities that lead most students to be successful in solving and recording addition problems that involve regrouping. Students also demonstrate understanding of subtraction with regrouping through manipulative-based activities that follow. However, even after direct instruction, very few can successfully make pencil-and-paper recordings that correlate to their manipulative solution processes.

Sample Discussion Issues

How do the *ticket* and *Race for Zero* activities differ?

What is the mathematical goal of the lesson and how well does it match the activities?

How do you connect manipulatives to abstract symbols?

Suggested Materials

Base ten materials, place value mats

Starter Problem

What thinking might children do as they figure out the answer to this problem? How might children record what they did?

$$100 - 5 - 2 - 4 - 1 - 3 = \underline{\hspace{1cm}}$$

Intentional Teaching: Comparing the *Ticket* and *Race to Zero* Activities

Both the *ticket* and *Race for Zero* activities focus mainly on subtracting from two-digit amounts, with successive amounts subtracted off. Also, both activities provide practice with regrouping (or trading) a ten for 10 ones, using a place value chart and base ten blocks.

But are these activities mathematically equivalent? One significant difference is that in the *ticket* activity students subtract two-digit numbers and in the *Race to Zero* activity they subtract one-digit numbers, the only kind on the die. Also, in the *ticket* activity, students work on the problems as if they were two single-digit problems: to subtract 32 tickets from 75 tickets, for example, students subtract the *ones* first, and then the *tens*. This may lead students to think of the answer as two separate numbers (*4* and *3*) instead of a whole quantity *(43)*.

Notice that in the *Race for Zero* game, when a small number is rolled on the die, students are likely to count back (*54* less *1* is *53*) rather than thinking of taking away ones and tens. This defeats the activity's purpose, which is to learn to subtract with regrouping. So, although the activities have similar surface features, they are not parallel and neither relates well to the teacher's goal. What might work better?

Intentional Goal of Learning to Subtract

Why are students so reluctant to write down the numbers in the *Race for Zero* game? Perhaps stopping to write down the numbers seems like a pointless activity that interferes with their goal to win the race to zero! The connection of the subcomponents of the activities to the mathematical goal of learning to subtract may not have been made explicit to them. Even when students do make a written record during the *Race for Zero* game, they miss an important step—the regrouping. They seem to think that the goal is to record what they see on the mat each time rather than to record the *process* that they are modeling. Also, students may just subtract mentally when they roll an easy number on the die (*90* minus *2* is *88*). If so, it may seem like *busy work* to write the problem down and do the *crossing out* to show regrouping.

Similarly, with the *ticket* activities, students may interpret the goal as doing several disjointed tasks, like *showing the number with blocks, making a trade*, and *take away some tens and some ones.* If tasks seem disjointed or unrelated to the overall goal of modeling a procedure for subtracting, students may miss the main point.

Language Links: Manipulatives to Symbols and Situations

The teacher is disappointed that the work students have done to record manipulative experiences hasn't transferred to solving problems that are *prewritten.* Why isn't there a transfer? Notice that the *ticket* and the *Race for Zero* activities involve successive subtraction. In other words, the *answer* for each problem becomes the starting number for the following problem. How would this be written? Students might naturally record the *Race for Zero* game as

$$100 - 3 - 5 - 3 - 2 - 4 \ldots = 0$$

How is this different from the common algorithm used for subtraction? The components of a subtraction problem (the minuend, subtrahend, and difference) are not explicitly illustrated in either activity.

Although the games develop fluency with trading, would it be better to use simple subtraction word problems rather than a game or activity as a context for learning the algorithm? Perhaps students could also benefit from practice making up their own subtraction situations for symbolic problems presented by the teacher.

Notes

5. Stop Sign or Equals Sign?

Our school is situated in an established blue-collar neighborhood, where I have been teaching for eight years. More than half of the second graders in my class are from non-European family backgrounds. In the past couple of years, I've been trying to help my second graders understand equations. The subject of equality recurs quite often. I have been fascinated by the students' thoughts on what the equals sign means and how many times it can occur in an equation. I have tried to create an environment that allows the students to direct the discussion. When needed, I ask questions or present examples to disturb their thinking. Then I allow them to struggle with their ideas until they came to a resolution.

January 23 Today Jimmy used the following problem as a way to make 98 (the number of days we have been in school):

$$(500 \times 1) - 500 + 97 = 97 + 1 = 98$$

Most students agreed with him, but Hal and Tony did not. Both said they didn't agree because you can't have two equals signs in one problem. Other students said you could. I asked, "What does the equals sign mean?"

Ed said, "Both sides have to be equal." Everyone seemed okay with that. Someone added that an equals sign is like a stop sign, and it has to be at the end.

So I asked, "What about $5 + 5 = 10$ and $10 = 5 + 5$, do these mean the same thing?" Everyone said yes because each side is 10. Several students concluded that the equal sign is like a stop sign, but it can go anywhere in the problem.

Tom chimed in that he agreed with Jimmy's problem and could explain why. He said that you do the first part and stop, do the middle part and stop, and then do the end.

© 2001 by the ExxonMobil Foundation

So I asked what the first part equaled, and everyone said 97. I asked what the middle part equaled, and they said 98, and the last part equaled 98. That caused some dissention in the ranks.

Someone said, "Well, just stop at 97, then do the + 1 later."

"Can you do it that way?" I challenged. "I thought you said the equals sign was the stop sign. Is the plus sign a stop sign, too?"

Several people said that you have to add the numbers together—you can't just stop. At this point, Hal asked to stop because he was getting a headache. Winnie asked if we could use a simpler problem to see the same thing.

So I said, "Well, what about $5 + 5 = 6 + 4$?"

Hal said, "No, add another equals sign so there are two, like Jimmy did."

So I wrote, $5 + 5 = 6 + 4 = 7 + 3$. There was a general murmur of agreement. Jimmy said, "Well, my problem is just like that. Why don't you agree with me?" A lightbulb went on, "Oh, mine is not equal on all sides!"

Tony said, "If you change the first 97 to a 98, I will agree with you."

So we changed it to read $500 \times 1 + 98 = 97 + 1 = 98$. But some people said, "No, it needs to be a 97." Someone asked if we had been working on this problem for 30 minutes yet.

"No, just 20 minutes," Hal said and mentioned his headache again. Jimmy asked us to leave the problem the way he had it originally, so we put it back.

Tony said, "Okay, but I don't agree anymore." We agreed to think about it some more.

March 29 Once again the subject of equality emerged in my classroom from a problem that Jimmy made up for our number of the day. Here it is:

$$138 - 0 + 0 + 1 - 1 = 137 + 1$$

Jimmy said, "Okay, does everyone agree with me?" Everyone agreed except Becky and Tom. I asked Tom why he didn't agree.

"Well, that first part equals 138, and Jimmy's problem says it equals 137." Lots of students tried to convince him that it wasn't what the problem meant.

Sandy commented, "Tom, it doesn't matter where the equals sign goes—each side is still the same in that problem."

© 2001 by the ExxonMobil Foundation

"No, the first part is 138, but the equals sign has 137 after it," Tom said.

Tony said, "Yeah, but it has 137 + 1, which is 138, just like the first part." Tom looked confused. I asked Becky why she disagreed. She said at first she disagreed for the same reason as Tom, but she changed her mind when she remembered what the class had decided.

When asked to explain, Becky said, "Ed said whatever was on each side of the equals sign had to be balanced, or be the same. So that problem is right."

Hal said that he understood why Tom disagreed, so I asked him to explain. "Tom thinks the equals sign is like a stop sign, and you only look at the 137 and not the + 1. I used to think that, but if both sides come out the same, it's equal. It doesn't matter where the equals sign is." Murmurs of agreement all around except from Tom, who looked frustrated. Everyone seemed ready to move on.

A student from a local university was observing that day and, at one point, went to explain to Tom why he should agree with the problem. I stopped him and said, "No, don't say anything. If Tom wants to disagree that is fine. He has to decide on his own what he thinks. We used to have a lot of people who disagreed." Later I took the college student aside and explained the history of the discussion.

April 2 Once again, the equality subject emerged. Here is the way the conversation went. Tony wrote:

$$141 + 1 - (1 \times 1) - 0 = 141 + 1 + 7 - 7 - 1 = 141$$

Jimmy quipped, "He learned from the master."

I asked, "Does everyone agree?" Even Tom agreed, which surprised me. "Tom, why do you agree with this one, but not with Jimmy's last week?"

"Because in this one the equals sign is at the end. In Jimmy's it was in the middle," he said.

Sandy noted, "But Tom, this one has an equals sign in the middle, too. It has two equals signs."

Tom responded that this one was okay because even though it had an equals sign in the middle, it had one at the end too."

"So," I said, "it's okay to have one in the middle if there is one at the end?" Tom said it was.

© 2001 by the ExxonMobil Foundation

Tony said, "All the parts still equal 141. I agree with Sandy that the equals sign can be anywhere." Seven raised their hands in agreement.

Jimmy asked, "What if the problem was just = 141?"

Someone said that wouldn't work because it wasn't a whole problem. Someone else said, "What about Nate's problem, does Tom agree with that?"

$$Nate's\ problem\ is\ 141 = 141$$

Tom said he agreed because 141 is the same thing as 141. I noted that Nate's equals sign was in the middle and asked, "How is Tony's problem different from Nate's?" Tony pointed out that he had two equals signs.

I erased the = 141 at the end of Tony's problem so that it read:

$$141 + 1 - (1 \times 1) - 0 = 141 + 1 + 7 - 7 - 1$$

When I asked how it was different, Sandy said it meant the same thing as Nate's, just longer.

"Tom, my first part of the problem is 141 and my second part is 141. It is like Nate's," said Tony.

Rethinking his ideas, Tom said, "Now I'm not sure."

April 16 Equality has become a favorite subject of our daily discussions. Now students bring up problems and ask to have them discussed. On Monday, the problem was:

$$147 + 1 - 1 = 147 + 2 - 2 = 147 + 1 + 4 - 1 - 4 = 147$$

Jimmy wasn't sure if he agreed. Anna wasn't sure either and asked, "Can there be three equals signs?"

Tom said, "I agree now that the equals sign can go anywhere, but all the parts have to be the same. In that problem, all the parts are the same."

Becky said it didn't seem quite right to her. "It is like it is two problems." When Becky was asked where the first problem ended, she said it ended after the first = 147.

"So the second problem is + 2 − 2 = 147?"

"Wait a minute, that can't be right. I see—the 147 goes with the + 2 − 2. Okay, I agree now."

On Tuesday, two students who had never given problems with more than one equals sign gave these:

© 2001 by the ExxonMobil Foundation

$$148 = 148 - 1 + 1 = 148$$

$$148 = 148 + 1 - 1 + 4 - 4 = 148$$

Everyone readily agreed and had no problem with either one. I was thrilled that they came to a consensus. My next challenge is to use their equations as a jumping off point for talking about some of the mathematical properties that they use intuitively in their equations. We have until the end of the school year to see where they can go from here!

Read and Reflect

Bay Area Mathematics Task Force. 1999. *A Mathematics Source Book for Elementary and Middle School Teachers: Key Concepts, Teaching Tips, and Learning Pitfalls.* Novato, CA: Arena Press.

Read pages 79 through 82 on algebraic thinking.

MacGregor, Mollie, and Kaye Stacey. 1999. "A Flying Start to Algebra." *Teaching Children Mathematics* 6 (2).

Read pages 78 through 85.

Witherspoon, Mary Lou. 1999. "And the Answer Is . . . Symbolic Literacy." *Teaching Children Mathematics* 5 (7).

Read pages 396 through 399.

© 2001 by the ExxonMobil Foundation

Stop Sign or Equals Sign?
Facilitator's Guide Notes

The second graders in this case participated in daily math discussions over the course of the school year. One of the most intriguing and enduring discussion topics was about the use of the equals sign. Several issues arose:

How many equals signs are allowed in a number sentence?

Is an equals sign like a stop sign?

Do the amounts on each side have to balance?

One student in particular disagreed with the majority consensus of his classmates, and the teacher allowed the disagreement to be resolved through discussion. The teacher and students used intentional examples to bolster their arguments. Finally, close to the end of the year, the student was convinced and changed his mind. This case highlights the teacher's skillful use of counterexamples and raises issues about the meaning of the equals sign.

Sample Discussion Issues

How might students' early work with number sentences prompt them to think of the equals sign as a stop sign?

What does the equals sign mean?

Is it correct to have more than one equals sign in an equation?

Suggested Materials

Counters, such as cubes or beans, and base ten blocks

Starter Problem

Which of the following examples show a mathematically legitimate use of the equals sign and which do not? Why?

$$5 + 5 = 10$$

$$10 = 6 + 4$$

$$5 + 5 = 10 - 4 = 6$$

$$5 + 5 = 6 + 4 = 7 + 3$$

Delineating the Boundaries of Equality: The Equals Sign

Everyone agreed that the equals sign means that both sides have to be equal or the same amount. However, some students thought the equals sign was also like a stop sign. In the example $138 - 0 + 0 + 1 - 1 = 137 + 1$, some students incorrectly thought 138 should follow the equals sign, instead of 137. In other words, they figured out the *answer* to $138 - 0 + 0 + 1 - 1$ and wrote it after the equals sign, or *stop sign*. This was troubling to students because although they agreed that $138 - 0 + 0 + 1 - 1$ and $137 + 1$ were equal, the number that followed the equals sign, 137, was in their opinion incorrect. This cognitive conflict fueled a debate that lasted several months.

What prompts students to think of the equals sign as a stop sign? Early work with basic facts teaches students to stop after the equals sign as you would at a stop sign, and write an answer. Students do lots of examples, like $3 + 4 = \_\_\_$ and $5 - 1 = \_\_\_$, but seldom do examples like $\_\_\_ = 3 + 5, 4 + 2 = \_\_\_ + 1$, or $2 + \_\_\_ = 6$.

The former examples have the equals sign and blank at the end of the equation rather than in the beginning or middle. The latter examples prompt students to think about making both sides equal, rather than putting the *answer* after the *stop sign*.

Notice that even the word *answer* has connotations that can be misleading. Students may think of an answer as the *result* after you do something. When a blank appears in the middle of an equation like $3 + \_\_\_ = 7$, students commonly write 10 in the blank because it is the *answer* to the addition problem $3 + 7$. Instead, if students learn to think about the blank as a *missing amount*, they will be building a foundation for thinking about equations in a future algebra course.

Students also struggled with an example that had two equals signs:

$$(500 \times 1) - 500 + 97 = 97 + 1 = 98$$

Again, those who interpreted the equals sign as a stop sign argued that the first part $(500 \times 1) - 500 + 97$ was equal to 97, and $97 + 1$ was equal to 98. However, the three amounts, $(500 \times 1) - 500 + 97$, $97 + 1$, and 98, were not the same. The first part is equal to 97, but the last two parts are equal to 98. It would be incorrect to write $(500 \times 1) - 500 + 97 = 97 + 1 = 98$.

Making Mathematical Properties Explicit

Tony's strategy for making an interesting equation was to start with a number and add zeros. The basis of his strategy was the *zero property*—adding zero to a number doesn't change the amount. For example, $141 + 0 = 141$. However, instead of adding a zero, he added expressions that were equal to zero, such as $1 - (1 \times 1)$. Students drew on the *multiplicative identity*—any number multiplied by 1 is equal to the same number—in their expressions. For example, in the preceding expression, Tony knew that (1×1) is 1, and Jimmy knew that 500×1 is 500 when he wrote $(500 \times 1) - 500$.

Several students also used the *additive inverse property*—the difference between the same two numbers is 0. For example, Tony purposely used $1 - 1$ and $7 - 7$ to make zeros in the equation $141 + 1 + 7 - 7 - 1 = 141$.

Intentional Teaching and Learning: Counterexamples

The teacher used intentional examples to challenge student thinking. When a student said the equals sign had to be at the end, the teacher chose two examples with simple numbers, $5 + 5 = 10$ and $10 = 5 + 5$, and asked if they meant the same thing. Since students quickly realized that these were the same, they reconsidered their assertion that the equals sign had to be at the end.

The teacher also contrasted the example $(500 \times 1) - 500 + 97 = 97 + 1 = 98$ with $5 + 5 = 6 + 4 = 7 + 3$. Both equations had two equals signs, but only two of the three parts of the first example were equivalent, while all three parts of the second example were equivalent. These examples led students to struggle with the conflict of using the equals sign as a stop sign or as a way to show balanced amounts on both sides.

Students also used carefully selected examples to argue their points; they argued that if the second and third equations were true, then the first equation must also be true.

$$141 + 1 - (1 \times 1) - 0 = 141 + 1 + 7 - 7 - 1 = 141$$

$$141 + 1 - (1 \times 1) - 0 = 141 + 1 + 7 - 7 - 1$$

$$141 = 141$$

Notes

6. Daily Equations

From the start of the school year my first graders begin talking and thinking about how numbers work. One of the things we do is a daily calendar routine that takes about 5 to 10 minutes of class time. All of my students enjoy this routine, even though they enter my class with a very wide range of experience and confidence levels.

In September, for our calendar activity, I ask the students to use their fingers to show the number of days we have been in school. So for the seventh day of school, for example, they might show three fingers on one hand and four fingers on the other. I ask them to explain how they decided to do it that way, and then we state the equation, 3 plus 4 equals 7. They quickly catch on and are eager to share different ways to show 7. I also show them how to use the calendar as a number line to check their number sentences.

In October, we start the numbers all over again, this time using the date, rather than the number of days of school, as the *target* number. We also begin to record equations. Each day, I prepare a large piece of paper on the bulletin board. I divide the paper into eight parts so that we can record six equations in vertical form and two in horizontal form. Then I write the number for today's date as a word at the top of the paper.

So for October 6, I ask students to show 6 with their fingers, we say the equation orally, and then I write down the equation. I write the equation either vertically or horizontally so that they learn it can be written either way. When they can think of no more equations, or when the paper is full, I stop recording. Here is an example of the equations they shared for October 6.

© 2003 by WestEd from *Number Sense and Operations in the Primary Grades*. Portsmouth, NH: Heinemann.

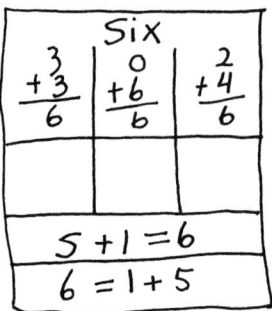

At first, only my more confident students participate. However, by mid-October I begin to call on students who are a little cautious. With time, even the more reluctant ones are ready to share. I keep a checklist to be sure that every child, including my English language learners and shyer students, get an opportunity to provide an equation.

At first, students come up with the equations randomly. Then one of the students notices the equations $4 + 2 = 6$ and $2 + 4 = 6$ and points out, "Hey those equations have the same numbers!" This provides an opportunity to talk about *turnarounds*—the commutative property. Of course, we then get lots more *turnarounds* when students generate their equations the following day.

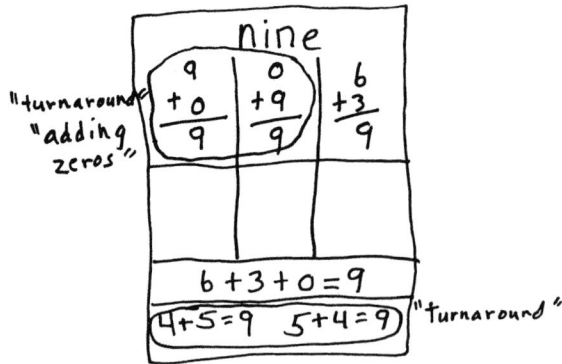

The same thing happens with the zero property. Someone brings up the example, 0 plus 9 is 9. I encourage further examples by saying, "Wow! Look what happens when you add zero."

Later in the year, they start bringing in equations involving more than two addends, such as $3 + 3 + 3 = 9$. This opens the door for a discussion of

© 2003 by WestEd from *Number Sense and Operations in the Primary Grades*. Portsmouth, NH: Heinemann.

repeated addition or multiplication. I say, "That is 3 three times, or three times 3." I draw a picture on the chart paper and record both number sentences.

$$3 + 3 + 3 = 9$$
$$3 \times 3 = 9$$

When a new month starts, the numbers for the dates are small again. Students quickly run out of interesting addition equations, so someone brings up subtraction. By February or March, they are recording equations on their own in a *Daily Equation* booklet made up of a few pieces of blank paper folded in half. I walk around looking at what they write as they work individually. This helps me assess what each student knows and still needs to discover. Also, when I spot an interesting equation, I ask the student to write it on a transparency to share during our class discussion.

It's great to see my students' progress from month to month. By April and May, they are really into patterns and even bring in place value ideas. They also seem to get a kick out of writing longer and more complicated equations. Some of Becky's equations form a pattern (see p. 70). For example, she realized that $7 - 3, 8 - 4, 9 - 5$, and $10 - 6$ are all equal to 4. Also, her understanding of place value is exhibited in the equations $20 + 4 = 24$, $10 + 10 + 4 = 24$, and $30 - 10 + 4 = 24$.

Joe's equations (p. 71) also showed a nice intuitive understanding of place value. He used multiples of ten very flexibly as the pages of his work show.

Kahlan was even using thousands and writing equations with two equals signs by the end of the year (see p. 72).

When I look back at my students' booklets, I can't help but be pleased with the progress they've each made. Their work amazes me. Yet, even at the end of the year, I see many incorrect equations. For example, I was thrilled that students like Allison realized that $5 + 1$ was the same as $1 + 5$, but disappointed that many like her mistakenly applied the same properties to subtraction (p. 72). Also, although some students were clear on adding and subtracting zeros, others like Marcus were confused (p. 70). Finally, I notice that overall, students are still reluctant to write equations like $8 = 4 + 4$ or $10 = 8 + 2$.

© 2003 by WestEd from *Number Sense and Operations in the Primary Grades*. Portsmouth, NH: Heinemann.

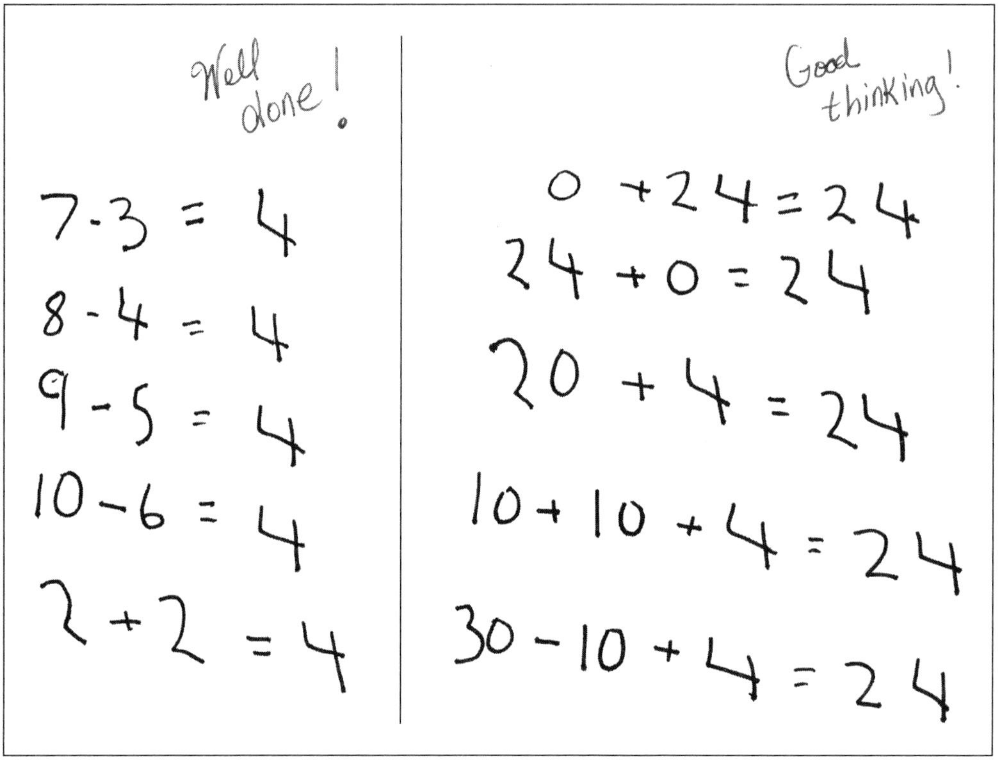

Well done!

$$7-3 = 4$$
$$8-4 = 4$$
$$9-5 = 4$$
$$10-6 = 4$$
$$2 + 2 = 4$$

Good thinking!

$$0 + 24 = 24$$
$$24 + 0 = 24$$
$$20 + 4 = 24$$
$$10 + 10 + 4 = 24$$
$$30 - 10 + 4 = 24$$

Becky's Equations

$$0 + 0 + 2 + 0 = 5 \text{ oops!}$$

$$10 - 0 - 0 - 0 - 0 - 0 = 5 \text{ oops!}$$

What happens when you add zero?

Marcus's Equations

© 2003 by WestEd from *Number Sense and Operations in the Primary Grades*. Portsmouth, NH: Heinemann.

$$10 + 10 + 6 = 26$$

$$10 + 10 + 10 - 10 + 6 = 26$$

$$10 + 10 + 10 + 10 - 10 - 10 + 6 = 26$$

$$10 + 10 + 10 + 10 + 10 + 10 + 10 - 50 + 6 = 26$$

$$100 - 74 = 26$$

$$10 + 10 + 7 = 27$$

$$60 - 40 + 7 = 27$$

$$30 - 3 = 27$$

$$7 + 10 + 10 = 27$$

$$70 - 50 + 7 = 27$$

$$80 - 60 + 7 = 27$$

$$90 - 70 + 7 = 27$$

Joe's Equations

© 2003 by WestEd from *Number Sense and Operations in the Primary Grades.* Portsmouth, NH: Heinemann.

Good Work!

$9000 - 9000 + 1 = 1$

$900,000 - 900,000 + 1 = 1$

$90,000 - 90,000 + 1 = 1$

~~$10,000,000 - 10,000,000 + 1 = 1$~~ oops!

$9000 - 9000 = 0 + 1 = 1$ ← check this one, too!

Kahlan's Equations

$5 + 1 = 6$ $1 + 5 = 6$

$0 + 6 = 6$ $6 + 0 = 6$

$10 - 4 = 6$ ~~$4 - 10 = 6$~~

$9 - 3 = 6$ ~~$3 - 9 = 6$~~

$2 - 4 = 6$ $4 + 2 = 6$

Allison's Equations

© 2003 by WestEd from *Number Sense and Operations in the Primary Grades*. Portsmouth, NH: Heinemann.

Read and Reflect

Bay Area Mathematics Task Force. 1999. *A Mathematics Source Book for Elementary and Middle School Teachers: Key Concepts, Teaching Tips, and Learning Pitfalls.* Novato, CA: Arena Press.
    Read pages 83 through 85 on algebraic thinking.

MacGregor, Mollie, and Kaye Stacey. 1999. "A Flying Start to Algebra." *Teaching Children Mathematics* 6 (2).
    Read pages 78 through 85.

Witherspoon, Mary Lou. 1999. "And the Answer Is . . . Symbolic Literacy." *Teaching Children Mathematics* 5 (7).
    Read pages 396 through 399.

© 2003 by WestEd from *Number Sense and Operations in the Primary Grades.* Portsmouth, NH: Heinemann.

Daily Equations
Facilitator's Guide Notes

The first graders in this case began the school year by learning different ways to show a *target* number using their fingers. Soon, they were learning to record equations to represent the sums shown on their fingers. By May, they were creating equations that used both addition and subtraction operations and larger numbers. In the process, the students noticed several mathematical properties, including the commutative property, the zero property, and the additive identity. They also used patterns to make their equations. The teacher was pleased with students' progress, but had concerns about how some students misapplied the mathematical properties.

Sample Discussion Issues

How might students learn that you can *turn around* addition but not subtraction problems?

What does the equals sign mean to a student who writes an equation like $9000 - 9000 = 0 + 1 = 1$?

How might a teacher capitalize on students' intuitive number sense?

Suggested Materials

Counters

Starter Problem

What mathematical properties are being used in the following equations? Do the properties apply the same way for both addition and subtraction?

$$3 + 0 = 3 \qquad\qquad 3 - 0 = 3$$
$$0 + 3 = 3 \qquad\qquad 3 - 3 = 0$$
$$1 + 2 = 2 + 1 \qquad\quad 2 - 2 + 3 = 4 - 1$$
$$2 + 1 = 0 + 3$$

Understanding the Boundaries of the Commutativity Property

The students in this case developed an intuitive understanding of *turn-arounds* through exploration, pattern discovery, and sharing. To solidify these ideas at a more formal level, they needed intentional guidance to know when the mathematical principle applies, when it doesn't, and why. For example, when students noticed that $4 + 2$ and $2 + 4$ had the same sum, the teacher could capitalize on this interesting observation by asking students to explain or show why. She might also ask students if *turnarounds* work with any numbers, no matter how big. Then the teacher might ask if $4 - 2$ was the same as $2 - 4$ and why or why not. This would give explicit attention to the boundaries of the principle. Students wouldn't mistakenly think that *turnarounds* always work, or that they only work with certain numbers. Unless the boundaries are clear, students may simply look at the surface features of a problem or of patterns and misapply the property, as Allison did in her year-end equations.

Understanding the Boundaries of the Zero Property

When one student gave the example $0 + 9 = 9$, the teacher encouraged further examples by saying, "Wow! Look what happens when you add zero." This could have also been an opportunity to draw explicit attention to the boundaries of the zero property. For example, students need to understand that if you add zero or subtract zero from another number, you get the same number, so $3 + 0 = 3$ and $3 - 0 = 3$. On the other hand, although $0 + 3$ is equal to 3, the parallel subtraction equation, $0 - 3$, is equal to negative 3, not positive 3. Some students may have overgeneralized the property because its boundaries were not made explicit. Marcus, who wrote $0 + 0 + 2 + 0 = 5$, and $10 - 0 - 0 - 0 - 0 - 0 = 5$ confused *adding zeros* with *adding ones*.

Another aspect of learning about zero is related to the additive inverse property. The additive inverse is the number that you add to another number to get a sum of zero. The additive inverse of a positive 3 is a negative 3, that is $(3 + -3 = 0)$. This is related to the idea that the difference between the same two numbers is zero $(3 - 3 = 0)$. Notice that Kahlan's equation, $9000 - 9000 = 0 + 1 = 1$ used both the additive inverse property $(9000 - 9000 = 0)$ and the zero property $(0 + 1 = 1)$.

Equality: Both Sides Have the Same Value

Most of the equations in this case have a series of addends on the left side of the equals sign and a single number, the *answer*, on the right side. Limiting students' experiences to this form of equation may have a negative impact on their future understanding of equivalency and equations, especially in algebra. The fundamental idea of an equation is that the amounts on both sides of the equals sign are equivalent or *equal*. So another way to pose the equations task is to ask students to write equations so that *both sides of the equation* have the same value as the *target number*. If the *target number* is 6, for example, one correct equation would be $3 + 3 = 2 + 4$, another $6 = 5 + 1$ and another $2 \times 3 = 6$. Collectively, these examples illustrate the broader concept of equations. Notice that Kahlan's equation ($9000 - 9000 = 0 + 1 = 1$) is incorrect because $9000 - 9000$ does not equal $0 + 1$ or 1.

In this classroom, students were asked to represent the *target number* with their fingers. One drawback to this approach is that students are actually representing an expression ($5 + 1$), not an equation ($5 + 1 = 6$). To represent an equation, rather than an expression, they would need to show 6 two different ways. For example, they might show 1 penny and 1 nickel in one hand and 6 pennies in the other hand. The amount of money in these two hands can be represented by the equation $5 + 1 = 6$. Or they might show 1 dime and 1 nickel in one hand and 3 nickels in the other to represent the equation $10 + 5 = 3 \times 5$.

Multiples of Ten and Place Value Concepts

Joe's various equations for 26 illustrate that he can compose and decompose numbers using a complex understanding of place value and multiples of ten. Notice, for example, that he added 7 tens together, then subtracted 50 and added 6 to get 26. It appears that he did this intentionally and that he understands that 50 is composed of 5 tens. Although it is not possible to know his thinking from the case, it is interesting to consider what it might have been. He may have thought, "I'll write down a bunch of tens, then take away all but 2 of them, because 2 tens is 20." Or perhaps he started by writing down $10 + 10 + 10 + 10 + 10 + 10 + 10$, then discovered that he had to take 5 tens away and then add back 6 to make 26. These equations also may illustrate an implicit understanding of associativity, or the idea that numbers can be grouped and combined in different ways. In other words, $(10 + 10 + 10 + 10 + 10 + 10 + 10) - 50 + 6$ represents the same value as $10 + 10 + (10 + 10 + 10 + 10 + 10 - 50) + 6$.

Notes

7. Diana

Diana is one of the most eager and hard-working first graders in my combination K–1 class. She really tries to do well in everything, especially in math. All of the students in my class get math in the morning when both classes are together. In addition, the 16 first graders, Diana included, spend another hour on math four afternoons a week.

During the first two months of school, the first grade students worked at math stations placed around the room, doing a variety of counting and addition activities with manipulatives. They also spent about a month practicing how to record number sentences to match the addition problems that they did with manipulatives.

I had thought that all of my first graders really understood what they were doing. Then I gave a four-page mathematics assessment at the end of the first quarter. The students did fine on the first three pages, which involved counting, comparing, and writing numbers for 5 to 19 pictured objects. On the fourth page, everyone also did well—until they got to the *10* abstract addition problems on the bottom half of the page.

Most of the children chose to do these problems without using manipulatives or pictorial representations. They either just *knew it*, or they used their fingers to count. Some students couldn't do the problems at all or made lots of errors. This was puzzling since everyone seemed to have completed the work successfully when we did the addition activities in our math stations.

Diana was one of those students who didn't seem to have the slightest idea what to do with the problems. When I noticed that she was having great difficulty, I gave her some cubes and told her that she could use them to help. Then I saw that she did not know how to use the cubes to do the problems either, so we did the first three problems together.

I don't understand why the activities we did during the weeks before the assessment didn't seem to carry over for some students. Here are some examples of the types of addition activities that Diana had previously been

© 2003 by WestEd from *Number Sense and Operations in the Primary Grades.* Portsmouth, NH: Heinemann.

Name **Diana**

Add.

1. 🌸🌸 🌸
   2 + 1 = *3*

2. 🐟🐟 🐟🐟 🐟
   1 + 4 = *5*

3. ⛵⛵ ⛵
   3 + 1 = *4*

4. ⭐⭐ ⭐⭐ ⭐
   2 + 3 = *5*

5. 2 + 2 = *4*

6. 0 + 3 = *3*

7. 1 + 1 = *2*

8. 1 + 3 = *3*

9. 1 + 0 = *1*

10. 3 + 2 = _____

11. 2 + 0 = _____

12. 4 + 1 = _____

13. 0 + 5 = _____

14. 1 + 2 = _____

able to complete in class when working at the stations. In one activity, the children shook a can and spilled out 3 counters that had been painted red on one side and blue on the other. Then they colored the pictures of the counters on the page to represent the real counters, and, finally, they recorded the numbers.

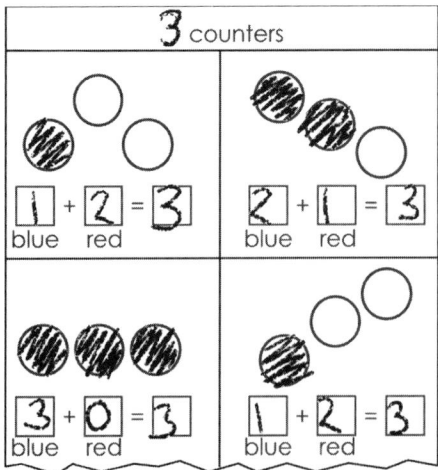

© 2003 by WestEd from *Number Sense and Operations in the Primary Grades*. Portsmouth, NH: Heinemann.

In another activity, the children shook out 4 painted counters. But instead of coloring the sets each time, they just placed the counters in the box that was drawn at the top of the worksheet. They then recorded number sentences to match each set of counters. In this way, they practiced finding different combinations for the same sum.

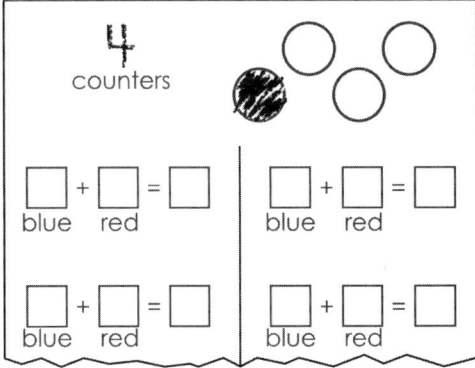

Then we learned how to play the *macaroni game* with a partner. I had my students start by putting four pieces of real macaroni on the left drawing of a cup and record *4 + 0 = 4*. Then they moved one piece of macaroni at a time over to the right *cup* and recorded the number sentence each time a piece was moved. They completed similar worksheets to practice finding combinations for other sums that were less than *10*.

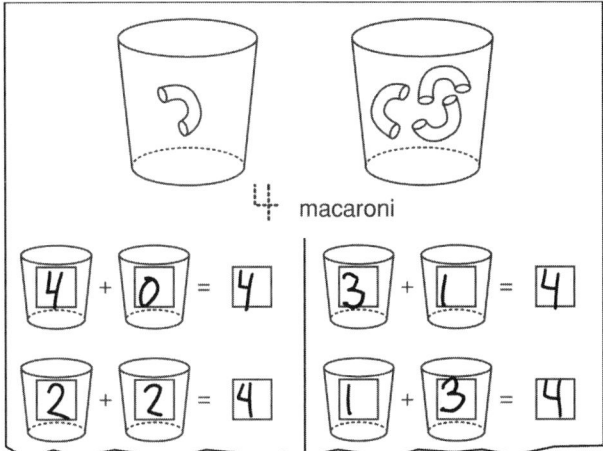

© 2003 by WestEd from *Number Sense and Operations in the Primary Grades*. Portsmouth, NH: Heinemann.

Finally, I wanted students to practice making up their own addition problems. They completed pages like the one shown in the following figure. I started by giving each child 5 teddy bears and asked them to put some in each *cave.* Then they wrote down the number sentence to show how many teddy bears there were in all.

After many addition experiences like these, when given a new addition activity, Diana would still ask me, "What do I do?" or "How do I do this?" Once I completed one or two examples with her, she was always able to continue on her own, and as far as I could tell, she was *getting it.* Obviously, though, from the results of the written assessment and her ongoing strategies, something didn't click for her—or for some other children either. It seemed like as long as I talked her through the examples, she could follow along and do the rest of the problems; but she didn't seem to be able to do the work without this one-on-one guidance.

Read and Reflect

Bay Area Mathematics Task Force. 1999. *A Mathematics Source Book for Elementary and Middle School Teachers: Key Concepts, Teaching Tips, and Learning Pitfalls.* Novato, CA: Arena Press.
Read pages 18 and 19 on number concepts and addition concepts.

© 2003 by WestEd from *Number Sense and Operations in the Primary Grades.* Portsmouth, NH: Heinemann.

Clements, Douglas H. 1999. "Subitizing: What Is It? Why Teach It?" *Teaching Children Mathematics* 5 (7).
   Read pages 400 through 405.

Kline, Kate. 1998. "Kindergarten Is More Than Counting." *Teaching Children Mathematics* 5 (2).
   Read pages 84 through 87.

Khisty, Lena Licón. 1995. "Making Inequality: Issues of Language and Meanings in Mathematics Teaching with Hispanic Students." In Walter G. Seceda, Elizabeth Fennema, and Lisa Byrd Adajian, eds., *New Directions for Equity in Mathematics Education.* New York: Cambridge University Press.
   Read pages 279 through 297.

© 2003 by WestEd from *Number Sense and Operations in the Primary Grades.* Portsmouth, NH: Heinemann.

Diana
Facilitator's Guide Notes

This case reveals a teacher's puzzlement over first graders' struggles with addition problems. The students have worked for more than three months in math stations using manipulatives, such as beans, cubes, and teddy bears, to practice solving and recording addition number sentences. Diana is one of the students who can do the problems after being led through a few examples, but can't figure out what to do on her own.

Sample Discussion Issues

How do we assess whether students' solutions are based on understanding or the superficial application of patterns?

What role does language play in understanding addition number sentences?

What are the benefits and limitations of always having each activity move from manipulatives to symbols?

How is the thinking required of students different for each of the addition activities in the lesson?

Suggested Materials

Beans, counters, linking cubes

Starter Problem

Say the following addition number sentence in your mind. What does each symbol mean? What might be difficult or confusing for a child?

$$2 + 5 = \_\_\_$$

Prompting Answers or Provoking Thought: Superficial Pattern Finding

Diana can complete her work successfully so long as she has help getting started. Might she be focused on finding the *secret* pattern needed to do each individual page, not realizing that the addition number sentences have meaning? Notice that the pages can easily be completed once a pattern is found. In the figure on p. 81 for example, she could count the red counters, count the blue counters, and write those numbers in the boxes directly below. Then she could write the number from the top of the page in the last blank. Although the teddy bear activity provided a word problem context, it was repetitive, and the students could easily fall into a superficial pattern-finding mode. Students may have benefited from oral problems using a variety of situations, such as "There were three birds and two more birds flew to the branch," or "Five children were on one side of a tug-of-war rope and 4 children were on the other side." Word problems are one way to help students associate language and meaningful situations with number sentences.

Language Links: Saying the Number Sentence

As adults, we might read the number sentence "3 + 2 = ____" as "Three things and two more are how many altogether?" Diana and other students may be missing the verbal language needed to interpret the number sentences on the assessment. What precautions need to be taken to ensure that students can read number sentences with meaning, especially in math station activities?

When reading number sentences, students must learn to ask themselves, "What is this telling me to do? What does this mean?" But some numeric symbols can be verbalized in more than one way. For example, the equals sign can be interpreted as *equals* or *is the same as*. Also, some symbols, such as the *blank*, may have multiple words attached to them, like *what number* or *how many*. Many teachers fail to make these subtleties explicit. What might help students understand what to do when they see the symbols?

Language Links: Manipulatives to Symbols and Back Again

How would children's thinking and internal language be different if they were given a number sentence and asked to represent it with objects rather than the reverse? Notice that in Diana's work prior to the assessment, she first did something with objects, and then she wrote the number sentence. Yet, on the assessment, Diana is unable to figure out how to do the addition problems even when given manipulatives. In order to use the counters to do the problems, Diana would have to be able to represent a number sentence with objects. What might help Diana understand how to think about the problems in either situation?

Addition Concepts: Action Versus Nonaction

Many of the activities in this case involve *nonaction* addition in a part–part–whole situation. For example, students count red counters (part) and blue counters (part) and then count how many counters there are all together (whole). But addition can also be represented as an action—a joining action. For example, 5 frogs (the initial set) might be joined by 2 more frogs. Joining situations readily invite students to use the counting shortcut of *counting on* from one set rather than counting *all* across both sets.

Why might it be important for students to have experience with addition in both action and nonaction situations? Students can easily make up their own action stories involving one set *joining* another set. It is more difficult for students to make up nonaction stories. Yet they need to develop the capability of associating both action and nonaction situations with the appropriate operation, in this case, addition.

Notes

8. Tallies and Coins

One of my major goals in math is to help my 31 first graders develop number sense. I try to do this in different ways, including using counting, place value activities, and, more recently, money activities. When I think back to how well my students have done on the counting and place value activities we've been doing all year, I'm a little puzzled about the difficulty they've been having with the later money activities.

Throughout the year, we've regularly practiced counting in many ways. One way that really helps with number sense, I think, is counting large numbers of objects, such as corn kernels, pumpkin seeds, and candy, into portion cups. To do this, the children count out 10 items per cup, leaving the extras on the table. Then, they count the tens and extras, saying, "10, 20, 30, 40, 41, 42, 43."

In addition, we have worked all year with the calendar. Our calendar time is a time to build math concepts, as well as a language practice time to build vocabulary for my students, including four ESL students. One thing we do during calendar time is count and record the number of days since we started school. We begin by reading the day and date for today and recording a new tally mark on a roll of adding machine paper to show one more day of school. Tallies provide opportunities to count groups of five, as well as tens and ones, building a foundation for counting money later. For example, if 18 days have passed since the beginning of school, we make a tally mark for each day, circling when we have made 10, like this:

Then the students say, "10 plus 5 is 15, and 3 more is 16, 17, 18," counting on as they read the tallies. I also model how to write and read the related addition sentence:

$$10 + 5 + 3 = 18$$

© 2003 by WestEd from *Number Sense and Operations in the Primary Grades*. Portsmouth, NH: Heinemann.

We also use drinking straws in cans labeled *tens* and *ones* to represent each day. After tallying our marks, we add a straw to our *place value* cans to represent one more day of school. When we get to 10 in the ones can, we *circle* the straws by putting a rubber band around them to make a bundle of 10 and put them into the tens can. We record the new number on an adding machine paper *number line.* Then, I ask them to say the number and to tell how many tens and how many ones there are.

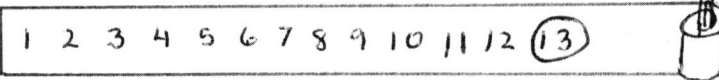

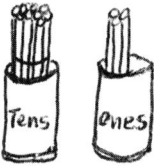

I frequently review the meaning of the numerals we are working with and the numbers they represent with the class. We compare the value of the tens and the ones. I do this by having them decide which place represents a larger amount. For example, I might ask, "Would you rather have 1 ten or 3 ones pieces of candy?" I think this is important to do throughout the year because it helps students begin to develop a firm sense of the values of the numbers in each place.

It is now late February. The students have experienced lots of these counting activities and are now studying money and the value of coins, including the penny, nickel, and dime. We have spent several days looking at real coins, discussing their attributes, and learning how much they are worth.

I spent the first few days asking them to count small numbers of coins of like value. I asked questions like, "How much money do you have altogether at your own table? At two tables? Three tables?" They came to agreement quickly and seemed to grasp this task easily.

The next step was to find the worth of collections of coins of unlike value. I limited this work to using pennies and nickels in combinations up to

© 2003 by WestEd from *Number Sense and Operations in the Primary Grades.* Portsmouth, NH: Heinemann.

10 cents. We first reviewed how a nickel could be exchanged for 5 pennies. Then I introduced how to count a nickel and pennies together. After some practice, it seemed they were ready for the next step. I gave each pair of students a cup of real nickels and pennies and instructed, "Show me as many ways as you can to make 8 cents." Several pairs of students then came to the overhead projector to show us how they did it.

George and Edwin showed a nickel and 3 pennies.

George did the counting, "This is 8 cents. The nickel is 5, 6, 7, 8 cents." He counted on as he touched each penny and wrote the value underneath.

Rachel and Gina came up and put 8 pennies on the overhead. They counted each penny to show that they also had 8 cents.

In another money activity, students went on a *shopping trip* for school supplies. On the overhead projector, I showed pictures of objects that we might buy. Then, making sure to pick combinations that cost less than a dime, I asked questions like, "If we bought an eraser for 3 cents and a pencil for 4 cents, how much money would we need? How do you know?" A typical response was, "We added 3 pennies and 4 pennies to get 7 cents."

Since the children understood how to do problems like these up to 10 cents, it seemed like they were ready to move on to counting with a dime, a nickel, and pennies. First, we reviewed how we could count a dime and some pennies. Then, I reminded them that we could count with a dime, a nickel, and some pennies just like we had counted the tallies.

© 2003 by WestEd from *Number Sense and Operations in the Primary Grades*. Portsmouth, NH: Heinemann.

So to count a dime, a nickel, and 2 pennies, we pointed to the coins and said, "10 and 5 is 15, 16, 17." They caught on quickly since this was so similar to counting tallies.

Then we went on another shopping trip. This time I showed the pictures of objects on the overhead and asked, "What two things would you like to buy?"

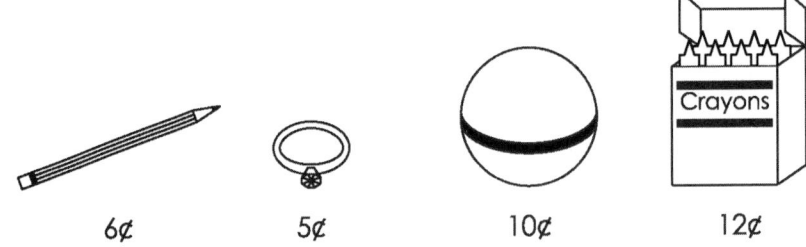

When one of the students said that he would like to buy a pencil for 6 cents and a ball for 10 cents, I asked everyone to work with their partner to show how much money it would take to *buy* the pencil and the ball. Most of the students had some difficulty figuring this problem out. But I noticed a few who were able to do it fairly quickly by using a dime, a nickel, and a penny to *pay* for the pencil and the ball.

Donna was one of the students who was successful on the problem, so I asked her to show the coins she used on the overhead. This is what she used.

© 2003 by WestEd from *Number Sense and Operations in the Primary Grades.* Portsmouth, NH: Heinemann.

When I called on students to tell how much the coins were worth, most of them couldn't tell me. This puzzled me because we had already done so much work counting 10s, 5s and 1s. Somehow when the order of the numbers changed, they had difficulty. If they had such good number sense, why couldn't they just add one plus five plus ten? In thinking back to our routines with the calendar and tallies, I realized we'd always counted quantities starting with the largest number (usually 5 or 10) first. Maybe if I had changed the routines every so often, they wouldn't have been thrown so easily.

Read and Reflect

Bay Area Mathematics Task Force. 1999. *A Mathematics Source Book for Elementary and Middle School Teachers: Key Concepts, Teaching Tips, and Learning Pitfalls.* Novato, CA: Arena Press.

    Read pages 14 through 17 on number concepts and addition concepts.

Geary, David. 1998. "Learning Mathematical Problem Solving." *Children's Mathematical Development.* Washington, DC: American Psychological Association.

    Read pages 95 through 130.

Fuson, Karen C., Laura Grandau, and Patricia A. Sugiyama. 2001. "Achievable Numerical Understandings for All Young Children." *Teaching Children Mathematics* 7 (9).

    Read pages 522 through 526.

© 2003 by WestEd from *Number Sense and Operations in the Primary Grades.* Portsmouth, NH: Heinemann.

Tallies and Coins
Facilitator's Guide Notes

This case presents us with many questions about developing the concept of number sense in young children. Students successfully complete several counting, place value, and coin activities. But when the students combine unlike coins to come up with a total, they run into apparent difficulty when the coins aren't in a descending order. The teacher questions whether the children's earlier work with coins and other counting tasks limited their thinking.

Sample Discussion Issues

Why do students easily recognize that 1 bundle and 3 straws is 13 straws, but have difficulty knowing the cost of two items, one 6 cents and the other 10 cents?

How were the counting tasks in the lesson similar and how were they different?

How might the place value and counting routines using the calendar and tallies help or hinder children's flexibility with numbers?

What are the benefits and limitations of prearranging quantities from greatest to least for children to count or add?

Suggested Materials

Play money, base ten blocks, straws, linking cubes

Starter Problem

Do the following tasks.

Count the coins in this order: 2 pennies, a nickel, and a dime. Now count them in this order: a dime, a nickel, and 2 pennies. How is your thinking different for each task?

Counting Coins: Order Makes a Difference

Many students couldn't count out the total amount when presented with a penny, a nickel, and a dime if the *smallest* coin was on the left, and the largest on the right. This collection of coins may have seemed *out of order,* based on their earlier experiences. They apparently didn't think of rearranging the coins to see if that would make it easier. Notice that in prior activities with tallies, straws, and coins, the values were arranged left to right with the largest on the left. For example, groups of ten straws were placed to the left and individual straws on the right. Possible advantages to this are that students might learn to order amounts from largest to smallest to count efficiently. They might also develop an intuitive sense of the base ten number system.

What are some possible disadvantages to prearranging quantities for students from greatest to least? One drawback is that students become so accustomed to having quantities ordered for them that they do not develop the flexibility to deal with values presented in different orders. It might be more advantageous to ask students to figure out for themselves the easiest way to arrive at a total.

Developing Flexibility: Counting More Than One Way

When the teacher asked students to tell how much a penny, a nickel, and a dime were worth, they had difficulty. This raised the question for the teacher about how well they had really developed their number sense in previous activities. Why couldn't they just add $1 + 5 + 10$?

Part of the problem may be that students had been taught to use one particular strategy for counting tallies, straws, and coins. They started with the larger value, and then counted on. For example, to count a nickel and three pennies, they began with five, then counted on, "6, 7, 8" cents. When they encountered numbers in a different order, the count-on strategy didn't work so well. For example, to count 2 pennies, 1 nickel, and 1 dime, it doesn't make sense to say 2, then count on 5 more, then count on 10 more. It would, however, result in the correct amount.

This points out the limitations of having students solve problems using only one strategy or of allowing routines to become too routine. Perhaps the teacher needed to *plant* a few challenges in the calendar activities to build a better number sense. For example, what if students had been asked to bundle straws into tens one day and sixes another day? They could discuss why it is easier with tens and how that relates to the number system. What if they looked at different ways of writing the number sentences for tallies so that they could see that 10 + 5 + 1 is the same as 1 + 5 + 10? What if they had used different strategies for counting the coins rather than always counting on?

Differences Between the Coin Tasks and the Tallies and Bundles

The teacher is puzzled about why the shopping activity is difficult for students in light of the work they have done building up to this activity. What are some possible reasons that students get stuck?

One possibility is that this is the first time they are given two numbers and asked to find a total. Another reason is that the shopping activity was presented as a multistep problem. First students were asked to represent the amounts in coins, and then they had to figure out how to count the total. But is it really necessary to use multiple steps? Students could have solved it by thinking "6 cents and 10 cents is like 6 straws and 10 straws so the total is 16." But they apparently didn't see the relationship to the earlier place value activities with straws. Although the students have experience *decomposing* 16 into two parts (10 and 6), they have had little practice starting with the numbers 6 and 10 and *composing* them into 16.

Notice also that coins are not visually composed of individual *ones* like tallies and straws are. Also, the bigger amount does not coincide with the size of the coins and may be confusing for some students. Yet, because of children's real-world experiences with money, the coins may carry more meaning than bundles of straws.

Counting: The Role of Five

What role might five play in children's development of a sense of quantities? One advantage to using five is that it is a small enough number for children to manipulate mentally. Children learn to skip count by fives quickly, and easily get that two fives makes ten. Additionally, five is a common component in counting money, since many coins are worth amounts that are multiples of five (nickels, dimes, quarters, and even half dollars). Another advantage is that using five provides children with some flexibility in the ways they arrive at a total and in the ways they break down a number into component parts. Using fives frequently might prevent children from becoming so focused on figuring out "how many tens and how many ones" that they lose sight of the total quantity.

Notes

9. How Many More?

In my inner-city, multicultural school district, 100 percent of the students receive free lunch. The district is participating in a grant in which teachers receive training to improve their mathematics and science instruction. My responsibility is to provide support to participating teachers. For the last four months, I have been co-teaching in Ms. B's first-grade class once a week. There are 22 students in the class; 18 of them speak Spanish as their primary language.

Ms. B and I have attended training to learn how to teach mathematics by having students develop and share their own strategies for solving word problems. We also learned how to make up a variety of word problems for students to solve and how to ask questions that would both help us understand their thinking and nudge it along. Ms. B's class works on word problems three or four periods a week.

On St. Patrick's Day, green footprints *mysteriously* appeared in two first-grade classrooms. The following problem was written on our board in English and Spanish:

> *There are ___ footprints in our room. There are ___ footprints in Ms. K's room. How many more footprints are there in Ms. K's room than in our room?*

After we read the problem aloud, I asked, "What do we need to know to solve this problem?" The students shouted together, "We need to know how many footprints there are!" First we counted the footprints in our classroom; then a delegation of four students was sent to count the footprints in Ms. K's classroom. Soon they returned to report finding 25 footprints. Now the entire problem read:

> *There are 19 footprints in our room. There are 25 footprints in Ms. K's room. How many more footprints are there in Ms. K's room than in our room?*

After we reread the problem in English and Spanish, students eagerly began working in pairs. Tubs of linking cubes were available. As is our

© 2003 by WestEd from *Number Sense and Operations in the Primary Grades.* Portsmouth, NH: Heinemann.

habit, the teacher and I passed among the students, talking with them about their work.

Dialogue with Mara and Teresa

My conversation with Mara and Teresa turned out to be almost identical to conversations with several other pairs of students. They could tell me how many footprints each classroom had, as well as which one had more. But even though I repeated the "How many more?" question several times, it was obvious they didn't understand what was being asked. After working with Mara and Teresa for a while on the footprint problem, I switched to a different problem.

TEACHER: How old are you, Mara?

MARA: 7

TEACHER: How old are you, Teresa?

TERESA: 6

TEACHER: Who is older?

MARA: I am.

TEACHER: How much older?

MARA: 7

TEACHER: How old is Teresa?

MARA: 6

TEACHER: How old are you?

MARA: 7

TEACHER: How much older is that?

MARA: 7

TEACHER: Are you a lot older or a little older?

MARA: A little.

TEACHER: How much older?

MARA: 7

TEACHER: Are you 5 days older?

© 2003 by WestEd from *Number Sense and Operations in the Primary Grades*. Portsmouth, NH: Heinemann.

MARA:	No.
TEACHER:	One day older?
MARA:	No.
TEACHER:	One year older?
MARA:	No.

Then I shifted our conversation back to the original problem.

TEACHER:	Who has more footprints?
MARA:	Ms. K.
TEACHER:	A lot more or a little more?
MARA:	A little.
TEACHER:	How many footprints do you have?
MARA:	19
TEACHER:	How many are you trying to get to?
MARA:	25
TEACHER:	How many more do you need to get there? How many more times does the little man need to walk in our room to be the same as Ms. K's room?

Mara wanted to use her fingers to count but seemed hesitant. After I let her know it was okay to use them, she counted up to 25 and responded, "6 more." Then she went back to her seat and recorded the following number sentence on her paper.

$$19 + 25 = 6$$

Dialogue with Thomas

Thomas brought up his paper to show me that he had gotten 30. I asked him how he got his answer.

THOMAS:	I counted both 25 and 19 and got 30.
TEACHER:	Did you count in your head or use the cubes?

© 2003 by WestEd from *Number Sense and Operations in the Primary Grades*. Portsmouth, NH: Heinemann.

THOMAS:	I used the cubes.
TEACHER:	Show me how you counted.

We walked back to his seat, and he recounted the cubes.

THOMAS:	I got 44. I counted 19 for our class and 25 for Ms. K and that's 44.
TEACHER:	Who has more footprints?
THOMAS:	Ms. K.
TEACHER:	A lot more or a little more?
THOMAS:	A little more.
TEACHER:	How many more?
THOMAS:	About 5.
TEACHER:	Go see if you can find out how many more there are.

A few minutes later he came back with an answer of 45.

Dialogue with Ruben and Nayeli

Ms. B. was working with Ruben and Nayeli. Each had used linking cubes to construct one train of *19* and one of *25*. They came up with the answer of *44*. Ms. B reread the problem to them at least three times before she went on to ask questions.

TEACHER:	¿Quién tiene más? (Who has more?)
RUBEN:	Ella tiene más. (She has more.) *He was pointing to the cubes representing Ms. K's room.*
TEACHER:	¿Cómo sabes que ella tiene más? (How do you know she has more?)
RUBEN:	Puedes ver. (You can see.)
TEACHER:	Múestrame como 25 son más que 19. (Show me how 25 are more than 19.)

These questions and answers were repeated four to five times. Then Ms. B held the 19 linked cubes and gestured toward the 25 linked cubes representing the footprints in Ms. K's classroom.

© 2003 by WestEd from *Number Sense and Operations in the Primary Grades.* Portsmouth, NH: Heinemann.

TEACHER: ¿Cuantas más tiene ella, que yo? (How many more does Ms. K have than me?)

NAYELI: 6 más. (6 more.) *She had lined up the two trains.*

TEACHER: ¿Cómo sabes que hay 6 más? (How do you know there are 6 more?)

NAYELI: Estos dos son iguales. Sobran 6. (These two match. There are 6 left over.) *She touched the trains where they met showing how there were 6 extra on the train with 25 cubes.*

RUBEN: Si le ponen 6 más son 25. (If you put 6 more, indicating adding 6 more cubes to the train with 19, there are 25.)

He used a count up to strategy.

RUBEN

M s K.
Porque Ms. K tiene 25 ella tiene 6 mas
Porque conte con los cubos

The students spent 35 minutes working on the footprint problem. Up to this point they have had at least 10 experiences with different types of comparison problems, and the results have been similar. There are always two or three students who are able to solve this type of problem, but they don't connect their strategies to other similar problems. With the footprint problem, 6 of the 22 students had correct answers, but only 3 could explain why. Eight students continued to add the two numbers together! As for the others, just from looking at their papers, it was hard to tell what they had done.

The teacher and I have talked about the difficulties students are having, and we attribute much of the problem to the fact that comparison problems

© 2003 by WestEd from *Number Sense and Operations in the Primary Grades.* Portsmouth, NH: Heinemann.

NAYELI

Yo conte ɸ dE el 25 y ES todo Lo QuE Sobro

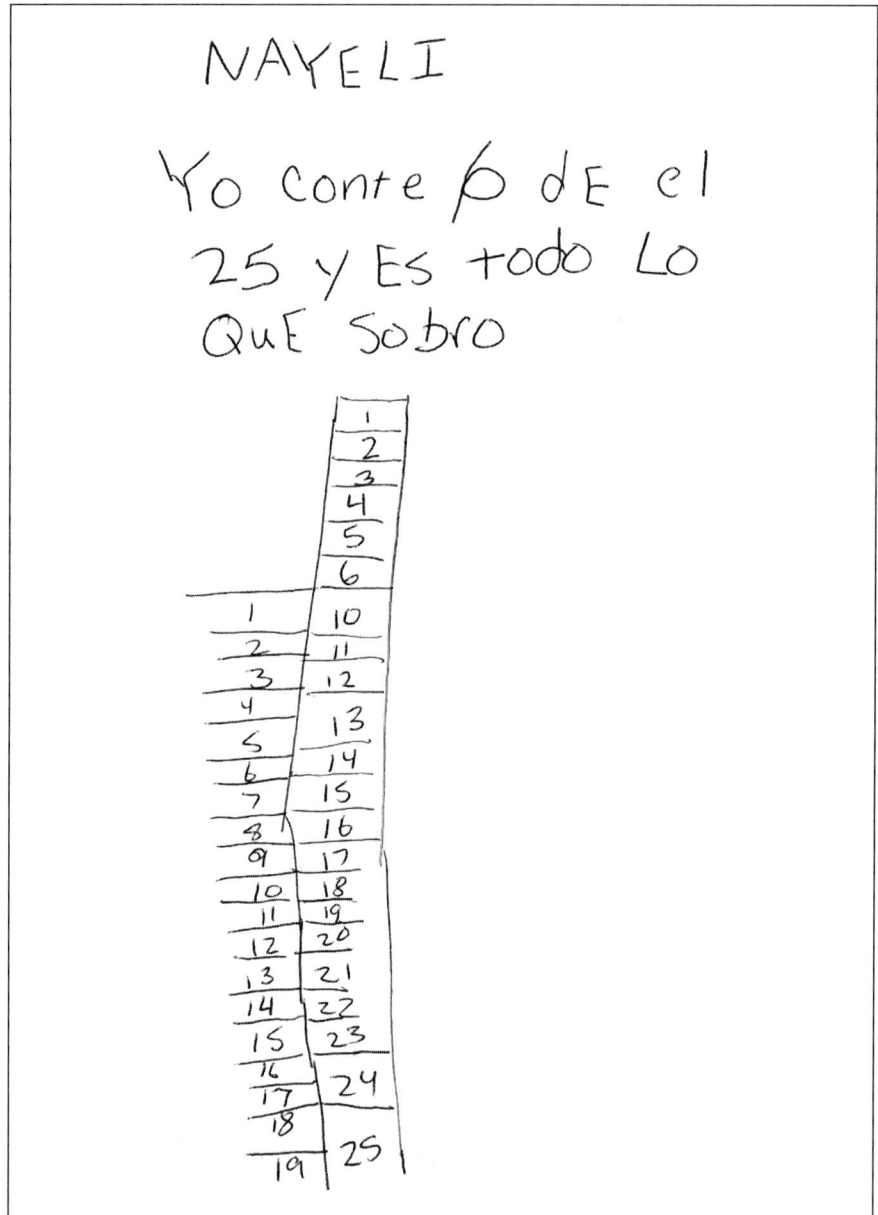

© 2003 by WestEd from *Number Sense and Operations in the Primary Grades.* Portsmouth, NH: Heinemann.

have no action that can be directly modeled. We also realized that the models for these problems can easily be confused with models for addition problems. Also, they are very different from models for *take-away* subtraction problems.

We've seen a lot of growth in students' problem-solving efforts and abilities during the past four months. The students are also more able to explain their strategies and do complex problems. Their confidence is strong and they enjoy doing math. Even so, we wonder whether there's something more that we could do to help them understand the meaning of "How many more?"

Read and Reflect

Bay Area Mathematics Task Force. 1999. *A Mathematics Source Book for Elementary and Middle School Teachers: Key Concepts, Teaching Tips, and Learning Pitfalls.* Novato, CA: Arena Press.
   Read pages 14 through 17 on number concepts and addition and subtraction concepts.

Geary, David. 1998. "Learning Mathematical Problem Solving." *Children's Mathematical Development.* Washington DC: American Psychological Association.
   Read pages 95 through 130.

Kamii, Constance, Barbara A. Lewis, and Bobbye M. Booker. 1998. "Instead of Teaching Missing Addends." *Teaching Children Mathematics* 4 (8).
   Read pages 458 through 461.

© 2003 by WestEd from *Number Sense and Operations in the Primary Grades.* Portsmouth, NH: Heinemann.

How Many More?
Facilitator's Guide Notes

A district support teacher works along side a first-grade teacher as the class tackles an engaging problem involving finding the difference between the number of footprints left in two classrooms. Only a small number of children solve the problem correctly. The teachers believe the reason this kind of problem is so difficult for students is that it has no action that can be directly modeled. They wonder what they can do to help students understand and solve problems that ask "How many more?"

Sample Discussion Issues

How do students interpret the phrase "How many more?"

What are the benefits and drawbacks of using *guiding questions* versus *leading questions?*

Why might a student write an addition number sentence for a *how many more* problem, even though they solved it by subtracting?

Suggested Materials
Linking cubes

Starter Problem

Draw a picture to illustrate this problem. Then write a number sentence for the problem. What about this problem might be confusing to a child?

There are 19 footprints in our room. There are 25 footprints in Ms. K's room. How many more footprints are in Ms. K's room than our room?

Subtraction Concepts: Why Are Comparison Word Problems Difficult?

The teachers say that *how many more* problems are more difficult because they "have no action that can be directly modeled." Whereas with *take-away* situations, students can physically model the act of removing something; with *how many more* situations, they must figure out the *relationship* between two quantities of things and find the difference.

The teachers also noted that models for comparison problems could be easily confused with models for addition problems. They most likely noticed that the students' models for comparison problems (like addition problems) began with two sets. Perhaps students were prompted to count both sets because they saw two sets. In contrast, take-away problems are modeled by starting with one set, then taking some away from that set.

A *how many more* situation can be modeled in different ways. Nayeli counts the *extras* that stick out after matching cubes from both sets. Ruben counts the difference by counting up from the smaller set to the larger set. Several other students are misled by the keywords *more*, or *más*, and decide, incorrectly, to add. Would it help students like Thomas or Mara to discuss the difference between a relative comparison question, such as *who is older* or *who has more*, and an absolute comparison question—*"How much older?"* or *"How many more?"* Or is it sufficient for first-grade students to understand relative comparisons without moving on to absolute comparisons?

Language Links: Word Problems to Symbols

This problem can be viewed as a *difference* ($25 - 19 =$ ___) or a missing addend problem ($19 +$ ___ $= 25$). After using her fingers to determine the correct solution of "6 more," Mara writes, "$19 + 25 = 6$." Students who think of the minus sign as *take-away* may not think to use it for a comparison problem. Perhaps Mara is thinking, "19 footprints plus how many more equals 25?" She may not know how to write the *missing addend* number sentence— "$19 +$ __ $= 25$"—which is correct. Is it unrealistic to expect first graders to write either the subtraction or the missing addend number sentence? Unless these ideas are modeled and discussed, students are likely to get stuck thinking the minus sign is associated only with *take-away.*

Students often have difficulty recording their thinking in drawings. Mara's drawing showed the 6 extra, but she did not line up the cubes accurately, and she skipped some of the numbering. What are the risks of assessing student understanding based entirely on their recordings?

Prompting Answers or Provoking Thought: The Role of Questions

Notice that the teacher begins her dialogue with Mara and Teresa by asking questions such as: "Who is older? A lot older or a little older? Are you 5 days older?" These questions about familiar ideas using smaller numbers can serve as temporary supports, or scaffolds, as students transfer their understandings to the footprints problem. However, when the teacher shifted to the footprints problem, she quickly went from questions that would scaffold, like those listed above, to a much more leading question: "How many more do you need to get there?" While this question prompts Mara to a correct answer, it probably will not help her learn to transform problems like these into similar questions for herself in the future. How might a teacher guide students into figuring out what questions to ask themselves when confronted with problems like these instead of leading them to an answer?

Notes

10. Why Is Subtraction More Difficult?

I've taught first grade for a number of years, and this year I have a particularly eager group of students. It's a multicultural group that includes four English as second language students.

One of my priorities in first grade is to help students develop their sense of quantities. Because of this, I always have my students begin the year by practicing counting. We count everything—students, crayons, whatever is around. Another priority is to help them understand the concepts of addition and subtraction, as well as to master the addition and subtraction facts. Over the years, I've noticed that the concept of addition seems to be much easier for students to understand than the concept of subtraction, and I've wondered why.

For addition, we focus on learning that *addition* means putting together sets of objects to make one larger set. We begin by using different manipulatives, such as beans, pencils, and linking cubes. My students especially like it when we combine groups of children or foods like crackers or pretzels.

We also spend a lot of time finding different combinations for a particular sum. For example, the students find different ways to put red and blue cubes together to show 5 cubes. We call these sets "5-trains." After many experiences putting together different kinds of objects, I have them write down the numbers in addition number sentences.

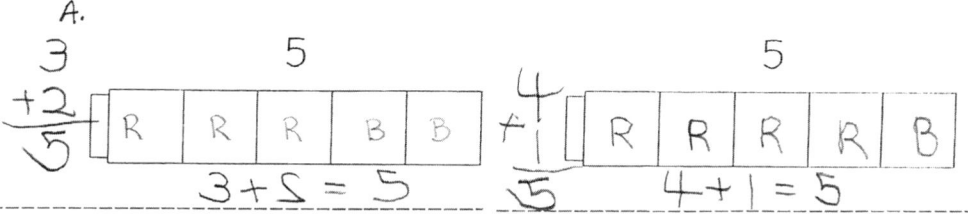

© 2003 by WestEd from *Number Sense and Operations in the Primary Grades*. Portsmouth, NH: Heinemann.

Recently, in response to concerns from parents and teachers about students not knowing their basic facts, our school implemented a math facts program for grades 1 through 6. The program is supposed to encourage students to memorize the facts, according to goals that are appropriate to their grade level. For example, during the first and second quarters, my first graders are expected to master addition facts whose sums are 5 or less. As part of the program, all students take a timed, three-minute basic facts test once a week. When they pass with 100 percent, they move on to the next level. So when my first graders get 100 percent on the addition facts, they move on to learn subtraction facts from 5 or less, ideally, during the third quarter.

I didn't give the first timed addition test until after my students had had a lot of practice using manipulatives to add, along with experience writing the corresponding addition number sentence. At that point, I thought they seemed ready to take the test. They were actually excited as we talked about how they should try to do as many of the 25 problems as they could in three minutes. On the first test, not everyone finished within the time limit, but with few exceptions, the students showed they knew their facts by having the correct answers on the problems that they did complete.

It really surprised me to see how much they liked to take these timed tests. After only two weeks of taking them, most of the students scored 100 percent on the addition tests and, according to the program, were ready to move on to subtraction facts.

Even though the whole class was successful with addition, my teaching experience told me that they weren't truly ready for lessons on subtraction, much less for the timed tests that would follow. But since they had completed the addition work, I decided we should go on to subtraction anyway.

To begin, I called five children from their seats to stand on the carpet and asked, "How many children do you see, class?" The students responded in unison, "Five!" I then asked Alicia and Nicky to sit down. "How many are left?" "Three," they all called out.

We continued oral practice with other groups of five students coming to the carpet and then subtracting some from the group. Then, we practiced with our *5-trains*. First, I told subtraction stories, and the children demonstrated with their *5-trains*. I recorded the number sentence on the board for the students to see and read. For example, for Alicia and Nicky's story, I would write the numbers as shown here.

© 2003 by WestEd from *Number Sense and Operations in the Primary Grades*. Portsmouth, NH: Heinemann.

$$5 - 2 = 3 \qquad \begin{array}{r} 5 \\ -2 \\ \hline 3 \end{array}$$

Soon they were ready to tell their own stories. "Who can tell a story using their train?" I asked. Alan answered, "There were five worms crawling on the ground. A bird came and ate one. How many worms are there?" Bobby said, "Four." I asked him to explain how he got that answer, and he said, "I took one cube away and have four left." As with the other stories, we wrote the number sentences on the board using both vertical and horizontal formats.

$$5 - 1 = 4 \qquad \begin{array}{r} 5 \\ -1 \\ \hline 4 \end{array}$$

After two or three days of practice, they demonstrated that they could link the work with the manipulatives to the written format for the subtraction problems. For example, they could make up stories to go with drawings of *5-trains* and then write number sentences to match.

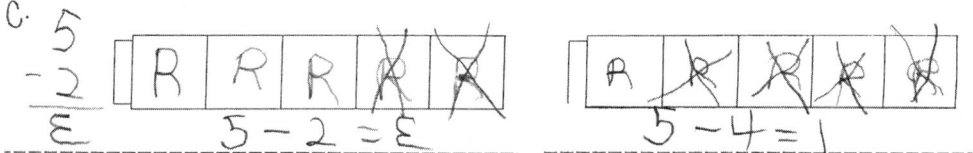

We then moved on to solving subtraction number sentences in horizontal and vertical form without first telling stories. I didn't encourage the use of manipulatives for these problems, so they used a variety of strategies, including counting their fingers and sometimes drawing pictures. For example, Fernando drew little ducks to solve the 5 − 1 problem.

© 2003 by WestEd from *Number Sense and Operations in the Primary Grades.* Portsmouth, NH: Heinemann.

$$5 - 1$$

$$5 - 1 = 4$$

After a few weeks of practicing with subtraction number sentences, without manipulatives or stories, I thought we must be ready for our first math facts subtraction test. I gave the test thinking that at least the strongest math students would do well.

When I collected the tests, I wasn't surprised that many were not finished. However, there were quite a few students who had finished the entire test! I was thrilled, until I began checking. Only Alan had gotten them all right. I had expected him to do well because he had quickly caught on to the addition facts and seemed to understand his subtraction facts. Most of the rest of the students had switched to addition on at least part of their tests, while others made a variety of mysterious errors.

Nicky's test was disappointing to me. I really thought she was *getting it* because she had been so verbal in telling subtraction stories and so active in making her *5-trains*. But on the test, it looks like she added! The only ones she got right were the ones in which she subtracted 0. It's almost as if she was focused more on the number in the whole train rather than on how many were left after some were taken away. Perhaps by always starting with the same number (5 students, 5 cubes, or 5 ducks), students focused on that as the answer. I wonder if it would be better to group the problems some other way instead of by number families.

The experiences I have had with my students this year lead me to believe that I have many students who have not internalized the meaning of addition and subtraction, especially in terms of written facts problems. This makes me hesitant to move on to higher sums and differences. They seem to *get* addition, but why do students always seem to have such difficulty with subtraction?

© 2003 by WestEd from *Number Sense and Operations in the Primary Grades*. Portsmouth, NH: Heinemann.

Subtraction Nicky

3	5	1	2	4
−2	−0	−1	−1	−2
⑤	5	②	④	⑥

5	4	5	3	2
−1	−0	−4	−0	−2
⑥	4	⑨	3	④

5	1	0	3	4
−5	−0	−0	−2	−1
⑩	1	0	②	②

2	5	4	3	4
−0	−2	−3	−1	−4
2	⑦	⑧	④	⑧

3	5	4	5	3
−3	−3	−2	−2	−0
②	④	②	①	3

Read and Reflect

Bay Area Mathematics Task Force. 1999. *A Mathematics Source Book for Elementary and Middle School Teachers: Key Concepts, Teaching Tips, and Learning Pitfalls.* Novato, CA: Arena Press.

Read pages 21 and 22 on addition and subtraction concepts.

Carroll, William M., and Denise Porter. 1997. "Invented Strategies Can Develop Meaningful Mathematical Procedures." *Teaching Children Mathematics* 4 (7).

Read pages 370 through 374.

© 2003 by WestEd from *Number Sense and Operations in the Primary Grades.* Portsmouth, NH: Heinemann.

Why Is Subtraction More Difficult?

Facilitator's Guide Notes

The first graders in this case are participating in a schoolwide math facts memorization program. They have had many experiences using manipulatives to represent and solve addition and subtraction problems. They have also made up their own story problems, used various strategies to solve them, and written the companion number sentences. After they did well on the addition test, the teacher reluctantly introduced subtraction. Then, after many successful experiences with subtraction, the students took the subtraction test. Unlike with the addition test, the results were disappointing. The teacher wonders why.

Sample Discussion Issues

What are the benefits and drawbacks of timed basic facts tests?

What is the role of understanding and practice in learning basic facts?

Why is subtraction more difficult?

Suggested Materials

Counters, linking cubes, beans

Starter Problem

Complete this subtraction number sentence:

$$5 - 4 = \underline{\hspace{1cm}}$$

If you didn't immediately know the answer, what are some ways that you might find it? What are some likely errors?

Basic Subtraction Facts: Developing Fluency

Should we assume that varied experiences over time will prepare students to recall number facts quickly? Some facts may be learned effortlessly through frequent use, others only through intentional memorization. Some facts can also be figured out quickly using counting strategies. The $8 - 7$ problem, for example, can be easily solved by either instantly recognizing that the difference is 1 or by counting up 1. The degree to which a particular counting strategy (i.e., count up from one number to another or count back from one number to another) is efficient depends on the individual math fact. For example, it wouldn't make sense to count back 7 to solve $8 - 7$. Note that the *count up* or *count back* strategies rely on understanding subtraction as a *difference* rather than a *take-away* situation.

Some students develop these strategies on their own but many will need instruction. What are the benefits and drawbacks of using quick counting strategies to recall facts? Note that students may become confused about which strategy to apply for which fact, especially under the pressure of a timed test. This may be, in part, why subtraction seems to be more difficult for students.

Subtraction Concepts: Fact Families

The teacher in this case places importance on having students learn to associate numbers in families. For example, they learn to associate 2, 3, and 5 so that they know that $2 + 3 = 5$ and $5 - 2 = 3$ and $5 - 3 = 2$. The *5-train* activities, for example, are used to develop a sense of number families for both addition and subtraction. Students use two colors of cubes to show combinations of 5 (i.e., $5 + 0$, $4 + 1$, $3 + 2$, $2 + 3$, and so on) and ways to subtract from 5 (i.e., $5 - 5$, $5 - 1$, $5 - 2$, $5 - 3$, and so on).

There may also be a downside to focusing mainly on fact families, since quick recall of number combinations may result in error patterns for some students. Nicky might be ignoring the subtraction sign, but it is also possible that she is simply recalling the wrong number family. It is easy for students to associate 5 and 4 with 9 instead of 1, especially if they learned addition first. How might this be avoided?

Basic Facts: Timed Tests

Many adults have unpleasant memories associated with taking timed tests. Do the benefits outweigh the drawbacks for learning basic facts under these conditions? Notice that the teacher in this case says how much her students looked forward to the timed tests for addition. Will students feel the same way about the subtraction tests? Is it possible to make the assessments less intimidating, yet ensure that they serve the same purpose? Giving plenty of time and/or fewer problems in the beginning is one way to lessen the pressure. Perhaps students might do only 5 or 10 problems in the beginning.

Another way is to show flash cards to the class and have students write down their responses. In this way, the teacher can give all students a little more time in the beginning and speed up as they become more proficient. The teacher might also immediately give the correct answers so that students get quick feedback. Otherwise, students can develop habitual incorrect responses to particular facts that can be very difficult to overcome.

Notes

11. Tina

My first graders spend about an hour every other day doing math through a problem-solving approach. On alternate days, we do hands-on activities as part of the district's adopted curriculum. We also do a daily math review sheet to add new skills weekly. By doing these different things, I feel that I am providing a good balance of activities for my students' different learning styles.

From the beginning of the year, Tina had shown herself to be quite mathematical in her thinking. During different math activities, both formal and informal, she would immediately have a correct response. For example, one day while the students were lining up, I presented this problem:

We had 11 students in line, but now we have 16.
How many kids just ran into line?

Tina immediately called out, "Five!" When asked how she knew this, she replied, "11 plus *5* is *16*, so *5* kids must have just come." She was very verbal and enthusiastic about mathematics and scenes like this took place frequently throughout the first part of the year.

Although my students had used math journals since the beginning of the year, most of the entries from the first semester were not related to problem-solving activities. The pages were filled with patterns, graphs, measurements, and other activities involving number sense. In January, I decided it was time for students to record their understanding of different math problems in their math logs. I asked questions such as, "A one-year-old has 4 teeth. A four-year-old has 12 teeth. How many more teeth does the four-year-old have?"

While other students drew things such as blocks, circles, mouths, and number lines, to illustrate their answers, Tina would show only a straight number sentence in her log, 4 + 8 = 12. She often wrote addition number sentences for problems like this when I was expecting subtraction. Yet, she knew the correct answer was 8, not 12.

© 2003 by WestEd from *Number Sense and Operations in the Primary Grades*. Portsmouth, NH: Heinemann.

This went on for several days, and both Tina and I were becoming frustrated. I wanted her to explain or make a drawing, but she just wanted to *do it*. Why could she write a number sentence and explain it orally, but couldn't draw, write, or symbolize how she got her answer in any other way?

I started insisting that all students put something on paper that showed their thinking. An entry that reflected only the number sentence from the problem was not allowed. Tina tried to comply with my wishes, but her written responses were not always easy for me to understand. For example, I presented the following problem one day:

There were 32 children on the bus.
Twenty-seven more students got on.
How many students were on the bus all together?

Most of the students drew a number line with marks denoting counting up, or they modeled the problem with blocks and then drew a picture of the blocks in their journals. Many of the students then wrote the number sentence 32 + 27 = 59. Meanwhile, this is what Tina wrote in her journal.

January 27

$$10 + 10 + 10 + 10 + 10 + 9 = 59$$

$$50$$

When questioned, Tina was not able to explain her number sentence. I continued to pose this type of problem, and Tina's responses continued in this vein for several days. Then, one day, when the children were lined up outside for recess, I engaged them in another one of our daily informal addition activities.

Mrs. C has 24 students in her line and Mrs. H has 23 in her line.
How many students are there all together?

© 2003 by WestEd from *Number Sense and Operations in the Primary Grades*. Portsmouth, NH: Heinemann.

Tina yelled out, "47!" When we got back to our room, I wrote the two numbers on the board and asked her to explain. She pointed to the 2 in the 23 and the 2 in the 24 and said, "Two tens and 2 tens is 4 tens, and 3 plus 4 equals 7, so my answer is 47."

Finally, it hit me! Tina was a step ahead of everyone else. She was already combining tens and ones in her head. I immediately drew a number tree on the board to illustrate what she had done.

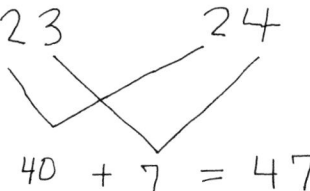

Later, in a one-on-one discussion, I modeled another math problem for her using a *tree*.

> *Myra has 11 smooth rocks and 24 bumpy rocks.*
> *How many rocks does she have all together?*

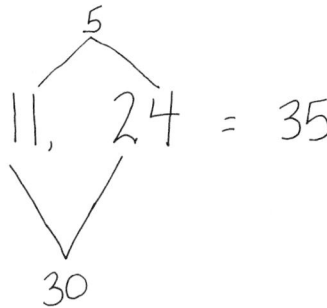

The very next day, we worked on finding totals for pairs of numbers, sometimes in word problems, sometimes just with *naked numbers*.

While many students only completed one or two, with no teacher help, Tina did four problems, each harder than the other. For the two problems with *easy* number pairs (7, 8 and 21, 9), she just knew her combinations. She solved them first and then tried to make the tree system work.

© 2003 by WestEd from *Number Sense and Operations in the Primary Grades.* Portsmouth, NH: Heinemann.

$$\overset{\text{7}}{V} + 8 = 15$$
$$7$$

$$\overset{21}{V} \\ 21 + 9 = 30$$

For the problems with the harder number pairs (20, 22 and 47, 56), she made the number tree to show her grouping of tens and then added the ones separately.

$$\overset{47, 56}{V} \\ 90 + 7 + 6 = \\ 103$$

$$\overset{20,22}{V} \\ 40 + 2 = 42$$

I began to realize that in recording her thinking process for the easier problems, Tina had struggled because she didn't really use direct modeling or strategies such as counting up or counting down to solve the problem. She just *knew* her math facts.

I know there are many other *Tinas* in my classroom, and in other classrooms. The systems that we ask them to use to solve math problems and record how they've done them doesn't always match up with the way they think about these problems. Since writing seems to help most of my kids clarify their ideas, I'm reluctant to give it up. So what do I do for students like Tina?

© 2003 by WestEd from *Number Sense and Operations in the Primary Grades*. Portsmouth, NH: Heinemann.

Read and Reflect

Bay Area Mathematics Task Force. 1999. *A Mathematics Source Book for Elementary and Middle School Teachers: Key Concepts, Teaching Tips, and Learning Pitfalls.* Novato, CA: Arena Press.
    Read pages 23 through 26 on place value concepts and multi-digit computation.

Chambers, Donald. 1996. "Direct Modeling and Invented Procedures: Building on Student's Informal Strategies." *Teaching Children Mathematics* 3 (2).
    Read pages 92 through 95.

© 2003 by WestEd from *Number Sense and Operations in the Primary Grades.* Portsmouth, NH: Heinemann.

Tina

Facilitator's Guide Notes

Tina is a student who *just knows* the answer but can't easily explain how she has figured it out. Because she is in a classroom that emphasizes oral and written explanations, she is struggling to meet the teacher's expectations. She adopts a method that her teacher has modeled, but she has difficulty applying it. This case raises the issue of how to help Tina make her thinking more explicit while not undermining her intuitive understanding.

Sample Discussion Issues

Why might asking students to prematurely record their thinking undermine intuitive understanding?

How does the action or relationship represented in a problem dictate how a student draws or thinks about it?

What are the benefits and drawbacks of encouraging students to use informal methods for adding before more formal procedures?

Suggested Materials

Base ten blocks, or other place value materials (optional)

Starter Problem

How might you figure out the answer to this problem if you hadn't been taught a written procedure for adding two-digit numbers? What are some ways you might show your thought process in writing?

Find the total of 47 and 56.

Prompting Answers or Provoking Thought: Undermining Understanding

It's sometimes easy to forget that students come to class already knowing a lot of mathematics. Some of their knowledge may have been actively developed at home, but much of it is more intuitive in nature. Tina appears to draw a lot on her intuitive knowledge. Although she can translate her thinking directly into a symbolic number sentence, she has difficulty explaining her thinking *process*. Her teacher wants students to be explicit about *how* they arrive at answers. Why might this be a good idea?

For many students, having to explain their thinking helps make their intuitive ideas more explicit and clear, and thus, more accessible to them in the future. Also, using pictures can be helpful because they stimulate ideas for solving a problem. Finally, when students share their thinking in public, they have opportunities to learn from each other.

Are there potential hazards to requiring students to explain their thought process for math problems? For students like Tina, explaining one's thinking may feel like *busy work* since she already knows the solution. She may feel pressure to make something up to please the teacher, even though it may not make sense to her. Notice that she latches onto the teacher's *number tree* explanation and tries to apply it to other problems. When asked to combine 47 and 56, she gets the correct answer, but has a discrepancy in her number sentence (refer to the art on p. 124, lower right). She combines 40 and 50, and then adds on 7 and 6 more. Her scribbled-out *branch* also indicates confusion about where to draw lines for the *tree*. Then she tries unsuccessfully to make the tree system work for 7 + 8.

It appears that there is a mismatch between Tina's ways of thinking and the teacher's modeling. Is Tina learning to reject her own understanding to please the teacher? Perhaps the teacher could work in partnership with Tina to develop a recording system that would not undermine her intuitive understanding. Another alternative is to wait until Tina faces problems that she can't solve intuitively—perhaps because they are more complicated or have larger numbers—and then help her develop more explicit reasoning about the problem.

Language Links: Missing-Addend Versus Difference

How does the action or relationship presented in a problem dictate how one might draw or think about it? When asked to find out how many more teeth the four-year-old has than the one-year-old, Tina writes an addition number sentence, $4 + 8 = 12$. How does she get the 8? She may think to herself, "Four and how many more makes 12?" This interpretation can be represented by the *missing-addend* number sentence, $4 + \_\_\_ = 12$; Tina somehow knows how to write the entire number sentence, but isn't able to explain her thinking process.

Is $12 - 4 = \_\_\_$ a legitimate number sentence for this problem? Both number sentences are correct, but represent different ways of thinking about the same problem. If you interpret the situation as finding the *difference* between 12 teeth and 4 teeth, a subtraction number sentence makes sense. Might the *missing-addend* and *difference* interpretations be represented in a different way pictorially, or would they look the same? Note that a *take-away* interpretation of this problem would not make sense, even though the same subtraction number sentence can represent either a *difference* or *take-away* situation.

Efficiency and Flexibility: Problem Solving before the Algorithm

The first graders in this classroom are solving problems using their own informal strategies and recording methods. How have students learned to combine numbers like 32 and 27 without first being introduced to the formal algorithm for adding? The teacher says that most of the students use direct modeling or counting-up and counting-back strategies to find their answers. So a student might count out 27 counters, count out 32 counters, and then count them all to find the total. Another student might find the answer by counting up 27 more from 32. Notice that Tina uses her informal understanding of multiples of ten and place value to help her figure out the answer $(10 + 10 + 10 + 10 + 10 + 9 = 59)$. She knows that the 3 in 32 means 3 tens and the 2 in 27 means 2 tens. She combines the 5 tens with 9 (7 ones plus 2 ones) to get 59.

Some would claim that two-digit addition problems involving regrouping are developmentally inappropriate for first-grade students, yet the students in this class are preparing to do such problems. Notice that even problems involving regrouping can be solved using modeling, counting strategies, and other informal methods such as the *tree* approach. What are the benefits and drawbacks of encouraging students to use informal methods for adding before they are asked to use or develop more formal written procedures?

Notes

12. Is Number Sense Enough?

I have been teaching first grade for four years. Math is a favorite subject of mine, and I think I have passed on my enthusiasm to my students. When they started last fall, they knew their shapes and how to pattern, but even the *highest* achieving students could not identify their numbers to ten. So I began the year working on number recognition and number sense, using number boards, counters, and games such as *War* to teach greater than and less than. Once they could identify and write their numbers all the way to one hundred and successfully compare numbers, we began to focus on addition and subtraction.

For three months we worked with one- and two-digit addition and subtraction problems without regrouping. I would give them a problem, and they would solve it using their own methods. Then we would talk with each other about how we solved the problems. They used counters, such as little blocks or their fingers, or counting strategies to figure out the answers in their heads.

I had started with addition and subtraction problems involving one-digit numbers and gradually increased the difficulty of the problems and the size of the numbers. Eventually, they were easily adding and subtracting two-digit numbers without regrouping and solving simple *missing number* problems like $4 - \square = 2$. Although I didn't put out base ten blocks, there were lots of activities that emphasized the meaning of place value. For example, students learned that 2 in the tens place means 2 tens or 20. And they understood how to add 2 tens and 1 ten to get 3 tens or 30. They had opportunities to work as a whole class, in small groups, and individually. By the end of three months, they all preferred not to use materials, doing the problems either mentally or with pencil and paper. They had also memorized most of the addition combinations and basic subtraction facts. They could

© 2003 by WestEd from *Number Sense and Operations in the Primary Grades*. Portsmouth, NH: Heinemann.

do any of the problems accurately and quickly on their own, using only pencil and paper.

$$
\begin{array}{r} 5 \\ +3 \\ \hline 8 \end{array}
\qquad
\begin{array}{r} 10 \\ +3 \\ \hline 13 \end{array}
\qquad
\begin{array}{r} 34 \\ +23 \\ \hline 57 \end{array}
\qquad
\begin{array}{r} 9 \\ +\boxed{3} \\ \hline 12 \end{array}
\qquad
\begin{array}{r} 3 \\ 5 \\ +2 \\ \hline 10 \end{array}
\qquad
\begin{array}{l} 3 + 4 = 7 \\ 5 + \boxed{3} = 8 \end{array}
$$

$$
\begin{array}{r} 5 \\ -3 \\ \hline 2 \end{array}
\qquad
\begin{array}{r} 12 \\ -3 \\ \hline 9 \end{array}
\qquad
\begin{array}{r} 34 \\ -23 \\ \hline 11 \end{array}
\qquad
\begin{array}{r} 9 \\ -\boxed{3} \\ \hline 6 \end{array}
\qquad
\begin{array}{l} 6 - 4 = 2 \\ 8 - \boxed{4} = 4 \end{array}
$$

With three months of school remaining, they seemed ready to move ahead, so I decided to introduce them to the second-grade curriculum—adding with regrouping. Just to see what would happen, I wrote 16 plus 17 on the board and asked them to solve it with no suggestions about what method to use. This is what they did.

$$
\begin{array}{r} 16 \\ +17 \\ \hline 213 \end{array}
$$

I then had them do the same problem again using individual counters that were on their desks. They all got 33, so I asked them to think about whether 213 or 33 made more sense as the answer and why. After a long discussion, they all agreed that 33 made more sense, but they didn't understand why they got 213 when they added using pencil and paper.

I suggested that we could use base ten blocks to solve the problem, and I showed them how. I modeled the traditional algorithm, where you start with the ones, then trade 10 ones for 1 ten if necessary, and then add the tens. We did many more addition problems like this one over the next three weeks, using manipulatives and emphasizing the language of *trading* 10 ones for 1 ten. Students recorded their answers, but weren't taught to show any regrouping notations.

They were very successful. After three weeks, I introduced the following written method to record the regrouping process. I was amazed to see

© 2003 by WestEd from *Number Sense and Operations in the Primary Grades*. Portsmouth, NH: Heinemann.

how quickly the students could apply the concepts that we had been practic-ing in our heads and with manipulatives to the paper-and-pencil method. They could all successfully explain the process, using place value language, and they could record the *trade* correctly.

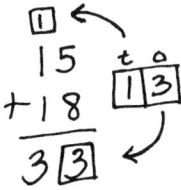

At the end of April, I noticed that some of my higher-achieving students had stopped using the algorithm I had taught, but were still getting the right answer. Looking more closely at their work, I noticed that they were not us-ing regrouping notations at all.

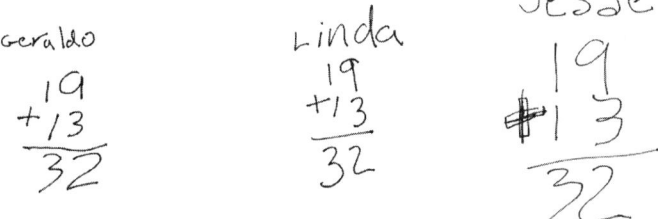

When I asked how they got their answers, they each had the same re-sponse, "I counted by tens first, and then added the ones." To add 19 and 13, for example, they counted, 10, 20, 29, 30, 31, 32.

At first I was very excited, since I thought their method showed that they understood place value and that they had developed a great number sense. Then I asked them if they could show me how they did it using the al-gorithm. This is when the confusion began. What they did is shown here.

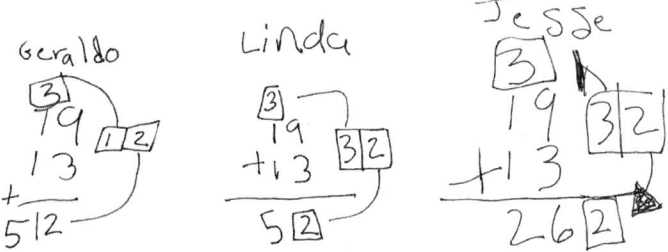

© 2003 by WestEd from *Number Sense and Operations in the Primary Grades*. Portsmouth, NH: Heinemann.

None of them could explain the process to me using any of the regrouping vocabulary that they had previously used very successfully. I worry that maybe they had memorized the pattern of the algorithm but didn't understand or forgot the meaning behind it. Should I go back and reteach the algorithm or let them rely on their number sense? What will happen when they have to add to the hundreds and thousands place next year, especially if they have multiple regroupings?

Read and Reflect

Bay Area Mathematics Task Force. 1999. *A Mathematics Source Book for Elementary and Middle School Teachers: Key Concepts, Teaching Tips, and Learning Pitfalls.* Novato, CA: Arena Press.

Read pages 23 through 26 on place value concepts and multi-digit computation.

Curcio, Francis R., and Sydney L. Schwartz. 1998. "There Are No Algorithms for Teaching Algorithms." *Teaching Children Mathematics* 5 (1).

Read pages 26 through 30.

Beishuizen, Meindert. 1993. "Mental Strategies and Materials or Models for Addition and Subtraction Up to 100 in Dutch Second Grades." *Journal of Research in Mathematics Education* 24 (4).

Read pages 294 through 323.

© 2003 by WestEd from *Number Sense and Operations in the Primary Grades.* Portsmouth, NH: Heinemann.

Is Number Sense Enough?
Facilitator's Guide Notes

The first graders in this case began the year with minimal number knowledge. But they made quick progress, soon tackling a variety of addition and subtraction problems that did not require regrouping, using counting strategies, manipulatives, and number sense. After three months, the students could add and subtract without using manipulatives and had memorized their basic addition and subtraction facts. Then the teacher used base ten blocks to introduce a pencil-and-paper algorithm for addition requiring regrouping. After lots of practice, all students could explain how the algorithm worked. Then the teacher realized that a few students were getting correct answers by adding mentally. Yet when asked to use the pencil-and-paper method they had *learned* in class, they were totally confused.

Sample Discussion Issues

What are the benefits and drawbacks of simultaneously teaching addition and subtraction from the beginning?

What are the benefits and limitations of always starting with the *ones* column when we do multi-digit addition or subtraction?

When and how do you introduce a pencil-and-paper algorithm?

How do you connect a student's number sense to an algorithm?

Suggested Materials

Counters such as cubes or beans, base ten blocks

Starter Problem

Add these numbers mentally. The first time, begin with the *ones* column. Then add them again, beginning with the *tens* column. Which method was easier and why?

$$33$$
$$+\ 28$$

Addition Concepts: Why Begin with the Ones Place?

Why do we start with the ones place when we do the traditional addition algorithm? Students probably need to understand why before they wade step by step through the algorithm. The main reason is that it is difficult to keep track of multiple regroupings with larger numbers. An algorithm that begins with the ones is needed especially when doing multiple regroupings for subtraction, but perhaps is needed less for addition. Consider how you would find the sum for 289 + 797 if you started with the hundreds place. You would begin by adding 200 and 700 to get 900. Then you would add 80 and 90 to get 170, which makes 1,070 when combined with the 900. Then you would add 16 more to get 1,086.

Students could also keep track of the multiple regroupings as they went along in this problem by writing down the partial sums of 900 + 170 + 16 and then adding. This method would work about as well as starting with the ones. However, doing a subtraction problem using this method (e.g., 724 − 538) is messy and not very efficient.

In the United States, the commonly used algorithms for both addition and subtraction begin with the ones place to simplify instruction. However, other countries use different algorithms for subtraction, some of which are considered easier and less prone to errors. Notice that 289 + 797 can easily be solved mentally, which raises the question about when an algorithm should be used.

Prompting Answers or Provoking Thought: Undermining Understanding

When mentally adding and subtracting one- and two-digit numbers, it really doesn't matter whether you begin with the tens or the ones place. As the students in this case demonstrated, if the numbers are relatively small, it makes perfect sense to start with the tens, even when regrouping is required. Yet, with the pencil-and-paper method most commonly used in the United States, students are strongly discouraged from starting with the tens column, even though number sense may suggest starting with the tens. Unfortunately, this can lead students to think that the number sense methods they have been using should be dropped entirely once pencil-and-paper procedures have been introduced. It is critical for students to retain their number sense so that they have a quick way to mentally check calculations.

It is common practice to introduce a pencil-and-paper algorithm when students first encounter two-digit number problems that require regrouping, even when the problems can easily be solved mentally. The intent is that by having students use the algorithm with simpler problems, even those they could otherwise solve in their heads, students will better understand the steps of the procedure when applying it to larger and more complex problems for which use of an algorithm is essential. The danger, however, is that students who can easily handle a problem using their number sense may not see the point of doing it the *hard way*, which may be how they view using the algorithm. Gradually, they may begin to think that mathematics isn't supposed to make sense and resort to memorizing algorithm patterns. The challenge to the teacher is to help students understand the importance of the roles both algorithms and number sense should play.

Developing Flexibility and Efficiency: The Value of a Good Algorithm

The teacher asks, "Should I go back and reteach the algorithm or let them rely on their number sense?" It is clear that number sense is a critical part of mathematical understanding. However, pencil-and-paper calculation methods also have an important purpose—providing reliable, relatively efficient methods to get correct answers, especially with large complex problems. That said, some algorithms are more effective than others. Several characteristics make a *good* (i.e., efficient) algorithm. First, it is consistent across many different problem types. For example, the algorithm should work with problems that don't require regrouping as well as with problems that do. Also, with an efficient algorithm, the same steps apply regardless of the number of addends or the number of digits in each addend. The steps in an algorithm are repetitive to keep errors to a minimum and to ensure that, with practice, the procedure can be applied *automatically.*

Number Sense: Linking Manipulatives and Algorithms

How do you help students make a transition from number sense and counting strategies to a written algorithm, especially when the algorithm runs counter to their invented strategies, such as starting with the ones rather than the tens? One common practice is to try to develop students' understanding of the algorithm by teaching the meaning of each step using manipulatives such as base ten blocks, while simultaneously showing students how to record the steps. The aim is to help students see how manipulations of base ten blocks to solve a problem can be mirrored in the algorithm's written procedures. One limitation to this approach, however, is that students like Geraldo, Linda, and Jesse sometimes memorize the patterns for doing the steps of the algorithm (e.g., one number goes up and one number goes down) without drawing on their number sense to help them work through the problem or check their solution.

What would happen if students were simply taught an algorithm and then asked to explain how they think it works? A possible outcome is that students would use number sense and reasoning skills to build their own understanding of the algorithm.

Notes

13. Closer to 500 than 400

For several weeks in my third-grade class, we used the Hundred Chart and Count by Tens Chart to explore tens and hundreds. Starting with the Hundred Chart, I asked my students questions such as, "What are some numbers that are closer to 20 than to 50? Greater than 50 but less than 75? What numbers have a 2 as one of their digits?"

I also asked them to find a particular number and tell how they found it. For example, when I asked students to find the number 34, one student said, "I looked on the top row and saw the 4, then I went down because I know 34 has a 4 in the ones place." Another student said, "I looked in the 30s and went across until I found 34."

When I asked, "Did you look at the top of the chart, the middle, or the bottom?" students responded, "I looked at the top because I know 30 is up there. The big numbers like 80 and 90 are on the bottom of the chart."

1	2	3	4	5	6	7	8	9	10
11	12	13	14	15	16	17	18	19	20
21	22	23	24	25	26	27	28	29	30
31	32	33	34	35	36	37	38	39	40
41	42	43	44	45	46	47	48	49	50
51	52	53	54	55	56	57	58	59	60
61	62	63	64	65	66	67	68	69	70
71	72	73	74	75	76	77	78	79	80
81	82	83	84	85	86	87	88	89	90
91	92	93	94	95	96	97	98	99	100

© Developmental Studies Center

They easily made observations about vertical, horizontal, and diagonal patterns. So based on their oral responses to my questions and on the fact that they had had experience with the Hundred Chart since first grade, I felt they had a solid understanding of the meaning of the numbers and the relationships among the numbers from 1 to 100.

I then had them explore the Count by Tens Chart. This chart is organized like the Hundred Chart, but increases horizontally by tens instead of ones. It increases vertically by hundreds instead of tens, beginning with 10 and ending with 1000.

After placing the Count by Tens Chart on the overhead projector, I asked students to look at it for about 30 seconds then turn to a neighbor and discuss at least one observation. Some of the comments I heard were:

"Every number has at least one zero."

"In the middle, there are 50s in the numbers."

"On the second row, if you take away the 1 in front of each number it goes 20, 30, 40."

"The numbers are getting bigger by tens."

10	20	30	40	50	60	70	80	90	100
110	120	130	140	150	160	170	180	190	200
210	220	230	240	250	260	270	280	290	300
310	320	330	340	350	360	370	380	390	400
410	420	430	440	450	460	470	480	490	500
510	520	530	540	550	560	570	580	590	600
610	620	630	640	650	660	670	680	690	700
710	720	730	740	750	760	770	780	790	800
810	820	830	840	850	860	870	880	890	900
910	920	930	940	950	960	970	980	990	1000

© Developmental Studies Center

"Most of the numbers are bigger than the numbers in the Hundred Chart."

"One side of the Hundred Chart and the top row of the Count by Tens Chart are the same. They both have 1 ten, 2 tens, 3 tens . . . all the way up to a hundred."

After these introductory activities, I introduced a game using two spinners, a record sheet, and a Count by Tens Chart for reference. In the game, pairs of students were to spin the two spinners (labeled 0–90 and 100–600, respectively) and add the resulting numbers mentally. Then, they were to decide where that sum would go on the record sheet, which included the categories, "Less than 200," "Closer to 500 than 400," "More than 200 but less than 400," "More than 600," and "Numbers that do not fit."

I introduced the game as a whole-class activity, with me spinning several numbers and the class computing the sums and discussing where each sum should be placed on the record sheet. Then I stated the academic and social goals of the lesson: (1) to find the sums of numbers on the spinners and decide where to place each sum on the record sheet, and (2) to share the work and explain their thinking to their partner.

To foster a classroom environment that promotes cooperative learning, I regularly have students work in pairs or in groups of four. I believe that children are more academically engaged and reflective when they are allowed

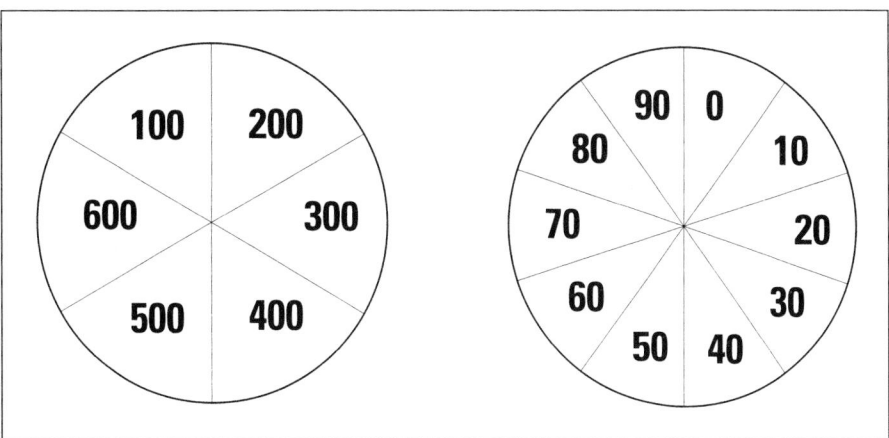

© Developmental Studies Center

Names _____

Is It Close?

Spin the two spinners, add the two numbers, and decide in which box to write the sum. Some of the sums can be written in more than one box.

Less than 200	More than 200 but less than 400
More than 600	**Closer to 500 than 400**

Numbers that do not fit

to explain their thinking and listen to the thinking of others. To get at both social and academic issues, I ask questions, such as:

How are you and your partner sharing the work?

How is your partner showing that he is listening to you?

© Developmental Studies Center

What are you doing to be a responsible learner?

What works well with your partner, and what does not work well with her?

How are you solving the problem?"

As partners began this activity, I walked around the room listening, conferring, and taking notes. I overheard several interesting conversations.

As Jose and Amelia spun 430, Jose said, "Let's put it in 'Numbers that do not fit.'" Amelia agreed.

I asked, "Why did you decide to put 430 in that box?"

Jose answered, "It isn't closer to 500 than 400. You would need 70 more to make 500, and you would only need to take 30 away from 430 to make 400. That's why it doesn't fit."

I turned to Amelia. "What do you think of what Jose just said?"

"Yeah, I agree," she said, pointing to the Count by Tens Chart. "'Numbers that do not fit' is right, because the number is closer to 400 than 500. If we just minus 30 away from 430, we'd get 400, but we would need to add 70 more to 430 to get 500."

These two students seemed to be using good number sense. They seemed capable of decomposing numbers, and they also explained their thinking and listened well to each other. As I moved around the room, however, I realized that the number relationships were not so clear for several other students. In fact, they seemed to find the record sheet quite confusing, especially the "Closer to 500 than 400" category. I also noticed that some students remained passive in their roles as learners. They seemed to simply accept their partner's thinking without becoming fully engaged in the activity.

For example, when I asked Anne and Terrance why they put 440 in the category "Closer to 500 than 400," Anne responded, "See, 440 is closer to 500 because it's on the same line as 500. It's not closer to 400 because 400 is on the line above."

Terrance agreed, but when I asked him, "Why do you agree?" he replied, "I just do."

Greg and Rashad were having a problem agreeing on the placement of 440 as well. Rashad seemed to be arguing for "Numbers that do not fit." However, Greg had another idea: "I think it should go into 'Closer to 500 than 400' because it's on the same line as 500."

© Developmental Studies Center

"But see," Rashad countered, "440 is 60 away from 500, but it's 40 away from 400. It's closer to 400 than 500."

Greg looked puzzled. Rashad continued using the chart to show Greg as he counted by tens. "Here, I'll count the numbers: 410, 420, 430, 440. That's 4 tens or 40. From 440 to 500 it's 450, 460, 470, 480, 490, 500. That's 6 tens or 60."

Greg was still unconvinced. Finally they compromised by agreeing to put 440 into both boxes. It was obvious to me that both Greg and Anne were focusing only on the proximity of the numbers on the chart rather than on the value or quantity. Despite Rashad's clear explanation, Greg was determined to hold on to his own thinking. I refrained from giving him another explanation, because I didn't think that would help. I wish I could have thought of a question to push his thinking at the time.

When we processed this activity as a class, several students expressed confusion about the term "closer to." Like Greg and Anne, they too seemed to be focusing on the proximity of the numbers on the Count by Tens Chart rather than the quantities. Thinking it might be helpful to return to the Hundred Chart, I placed it on the overhead projector. I was surprised to hear students express similar confusion about the numbers on that chart.

"Forty is closer to 50 than 48 is," said Lou. "See, 40 is right above 50 and 48 is two boxes away." Classmates tried to help with explanations like "40 is 10 away from 50, but 48 is only 2 away from 50," but Lou was unconvinced.

Anita said that 38 is closer to 30 than 40. When asked why, she explained, "Thirty-eight is in the 30s, not the 40s." Even after our discussions, some students were still unable to see the number relationships that I thought these lessons would have made clear.

Thinking back on this particular lesson, I realized that although students could find patterns in the chart, they didn't necessarily understand the relationships. I wondered what would have happened if we had cut up the Hundred Chart and made a long number line so that they could have seen the distance between the numbers more easily. I'm not sure they realized how the chart was related to counting.

I was also concerned about the passive learners who agreed with their partner's explanations without fully understanding the thinking; and I worried about students who disagreed, but never had a chance to talk about their disagreements. I continue to struggle with ways to support all my students both mathematically and socially.

© Developmental Studies Center

Read and Reflect

Bay Area Mathematics Task Force. 1999. *A Mathematics Source Book for Elementary and Middle School Teachers: Key Concepts, Teaching Tips, and Learning Pitfalls.* Novato, CA: Arena Press.

Read pages 23 through 26 on place value concepts and multi-digit computation.

Bishop, Alan J. 2001. "What Values Do You Teach When You Teach Mathematics?" *Teaching Children Mathematics* 7 (6).

Read pages 346 through 349.

Klein, Anton S., Meindert Beishuizen, and Andri Treffers. 1998. "The Empty Number Line in Dutch Second Grades: Realistic Versus Gradual Program Design." *Journal for Research in Mathematics Education* 29 (4).

Read pages 443 through 464.

© Developmental Studies Center

Closer to 500 than 400
Facilitator's Guide Notes

This case describes lessons taught in a third-grade class using a Hundred Chart and a Count by Tens Chart. The teacher discovered that, although her students observed many patterns in the charts, they were not necessarily connecting the patterns to mathematical ideas, such as relative magnitude or place value. So, when students were asked to sort numbers into categories, such as "closer to 500 than 400," some students interpreted the task differently from others. This case stimulates discussion about how to use the charts so that students gain a better understanding of key mathematical concepts. This case also raises questions about asking partners to agree on solutions when, in some cases, one partner may be incorrect and cannot be persuaded otherwise.

Sample Discussion Issues

What does *closer to* mean to students?

How do we assess whether students' solutions are based on understanding or on superficial pattern finding?

What are the benefits and limitations of using a Hundred Chart or a Count by Tens Chart?

How can you encourage both cooperation and correct mathematics?

Suggested Materials

Hundred Chart, Count by Ten Chart, number lines, and base ten blocks

Starter Problem

Use materials or make a diagram to show that 440 is closer to 400 than 500.

Language Confusion: Closer in Proximity or Closer in Value?

What do we mean by *closer?* Some third graders in this case focused on the *proximity* of the numbers in the chart rather than on comparing their values, or *relative magnitude.* For example, Greg said that 440 is closer to 400 than 500 because "it's on the same line." Similarly, Lou said that 40 is closer to 50 than 48 because "40 is right above 50 and 48 is 2 boxes away." Anita, basing her thinking on misleading cues, thought that 38 was closer to 30 than 40, since "38 is in the 30s."

Students may need explicit instruction to understand the distinction between *closer to* as it applies to proximity and *closer to* as it applies to quantity or value. The difference between these interpretations may be confusing, especially for second language learners. It may help if students discuss the meaning of *closer to* in context such as: "If your sister is 18, is she closer to being 10 or 20 years old?" Then, this situation could be related back to the chart.

Intentional Instruction: Relative Magnitude

Although the Hundred Chart and the Count by Tens Chart are beneficial for helping students see patterns in the base ten number system, they can also obscure mathematical concepts unless instruction addresses these issues more explicitly. Many of the early activities outlined in this case can be successfully accomplished by looking at the visual characteristics of the chart or by simple counting. Children may be *learning the chart* rather than the mathematics implicit in the task.

Although the stated goal of the last lesson in the case was to find sums and place them on the record sheet, the teacher's real and unstated goal for her students was to have her students compare and categorize the sums by their relative magnitude. While students were not asked to round numbers to the hundreds, the activity provided an excellent foundation for this skill. Although students like Jose and Amelia had a strong sense of how great 430 was compared to 500 and 400, this concept was not so clear to other students. What did these students need to understand? One of the goals for students should be to understand the relationship between (1) the value or magnitude of a number, (2) counting (by ones, tens, or hundreds), and (3) representations such as the number chart or the number line.

Perhaps, as the teacher suggested, another model, such as cutting up the Hundred Chart to make a linear number line, would help students relate these ideas. Then students could visually compare the distances between the numbers. Some students might need to see all of the numbers, not just 400, 410, 420, 430, and so on. They might need to actually write out the whole number line with 400, 401, 402, 403, 404, and so on up to 500. Students could be asked how far it is from 400 to 430 or from 430 to 500 and to explain how they know using the models and counting. Eventually, they could develop a *mental* number line to solve these kinds of problems.

Cooperation and Mathematical Integrity

Rashad and Greg each came up with a different answer to the question of how to categorize 440. After following the teacher's directions to explain their thinking to each other, they still didn't agree. So, they compromised by putting 440 into two categories, yet Greg's understanding of the mathematics was incorrect. The teacher was looking for a question that might push his thinking, without usurping the responsibility she had bestowed on each pair of students to iron out the mathematics themselves.

One possible explanation of Greg's erroneous conclusion is that he simply misinterpreted the words *closer to*. Perhaps the teacher could have asked both students to explain how they defined *closer to* so that they could grapple with their different interpretations and perhaps come to a common understanding.

Another possible explanation is that Greg's number sense foundation was not solid enough to support sophisticated reasoning about the numbers. If this was the case, the teacher could ask about a more obvious number relationship, such as "Would 490 be closer to 400 or 500?" Over time, such questions might help Greg see the flaw in his thinking and begin to build a foundation for reasoning about less obvious relationships.

A long-term goal might be to help students learn how to ask questions about each other's thinking rather than just explaining their own thinking and passively accepting others' explanations. For example, Greg might have asked Rashad, "I still don't understand what you mean when you say that 440 is 60 away from 500. Could you explain it a different way?"

Notes

14. Dollars and Cents Confusions

As part of a pilot study at our school, students are assigned to cross-grade math groups based on their number sense, which we determine through one-on-one interviews and a review of their work samples. Our hope is that in grouping students who have similar readiness for a particular math topic, we can accelerate their mathematical development by having them do problem-solving, mental computation, and math games targeted at just the right level. Students can be moved from one group to another at any time based on their progress. My own group of 18 third- and fourth-grade students, 5 of whom have special needs, consists of the least able mathematicians at these grade levels.

For several weeks early this year, I worked with my students on three different ways to think about numbers. First, I had them work with pairs of numbers that combine to make 10, then extend these pairs into multiples of 10 that combine to make 100—for example, $3 + 7 = 10$ and $30 + 70 = 100$. Second, I focused students on seeing ways to make 10 or multiples of 10 when adding a string of numbers. For example, in $16 + 4 + 7 + 13 = \_\_\_$, students could think $16 + 4 = 20, 7 + 13 = 20$, and $20 \times 2 = 40$. Third, I presented students with problems involving money, because I thought money illustrated the base ten number system well, and many of my students were familiar with money from their lives outside of school. It also gave them opportunities to apply both addition and multiplication skills.

One Friday, I wanted to assess my recent lessons by asking students to solve a problem adding combinations of coins. I purposely chose coin combinations that, when added, made sums of 10, multiples of 10, and 100. I hoped students would use some of the computation strategies they had learned over the preceding weeks. I presented the following problem:

Elise and Kim went to the school gift shop. They had 4 quarters, 6 dimes, and 8 nickels. How much money did they have together?

© Developmental Studies Center

By this point in the year, students knew that they could use numbers, tallies, pictures, or words to show their thinking. As they worked on the problem, I handed out transparencies and overhead pens to those who wanted to show and explain their solution later during the whole-class discussion.

As individual students finished solving the problem, they moved to the math game center to play games until everyone was finished. During the last 20 minutes of class, I called the students together to share their strategies. I had taught them to show agreement or disagreement with each other by showing thumbs up or thumbs down, or by saying "agree" or "disagree."

Rosalie volunteered to go first. She placed her work on the overhead projector.

"I knew 4 quarters made a dollar," she said, pointing to 4 circles with the number 25 written on each. "Then I put 6 dimes." She pointed to six smaller circles with the number 10 written on each.

"How much is that?" I asked.

"Sixty cents," she said.

"How did you get 60?"

"I counted on my fingers." She held up six fingers and counted by tens. I nodded for her to go on. "Then I knew nickels are 15, so I made nickels like this." She pointed to five circles, each labeled with the number 15.

© Developmental Studies Center

"Disagree!" said Elise. "You got to put 5 on the nickels, not 15. And you got to make 8 of them."

"Oops! I mean nickels are 5," Rosalie said, and she crossed out the 15s and rewrote them as 5s.

"What made you change your mind?" I asked.

"I got mixed up when I was counting," she responded.

Steven raised his hand and offered, "Now all you do is multiply 5 × 8 and you get 40."

Rosalie counted by 5s just to make sure, then wrote the number 40 next to her nickels. "Okay," she said. "Now I put 1 dollar and 60 and 40 together and got . . . 11 dollars."

"Disagree!" Raymond shouted, showing a thumbs down.

"Why?" I asked.

"Because 4 quarters makes 1 dollar, then 60 cents and 40 cents more aren't 11 dollars. It's only 1 more dollar, because 1 dollar and 1 dollar makes 2 dollars."

I asked him to bring his work to the overhead projector. Like Rosalie, he had used a combination of pictures and numbers to solve the problem. However, in writing his answer, he used the decimal point to indicate he understood the place value of the dollars and cents.

"What do you think, class?" I asked. Some students showed a thumbs up, while others gave a thumbs down.

© Developmental Studies Center

Melinda raised her hand. "I got a different answer," she said, bringing her work to the overhead projector. "Four quarters is $1.00. Six dimes makes $1.60. Forty more makes $5.60."

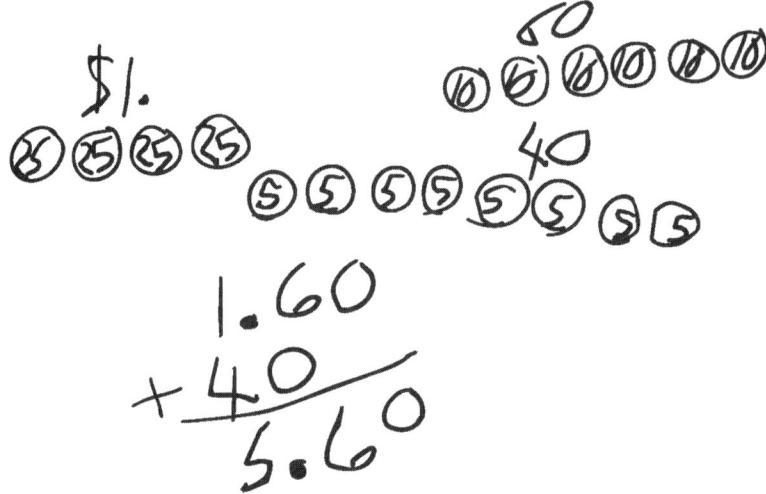

"I disagree with everyone so far," said Elise. She brought her work to the overhead projector. "I almost agree with Raymond, but I got 2 dollars plus 1 cent because 40 plus 1 equals 41, plus 60 equals 101. Plus 4 quarters makes it . . . 200 plus 1 cent."

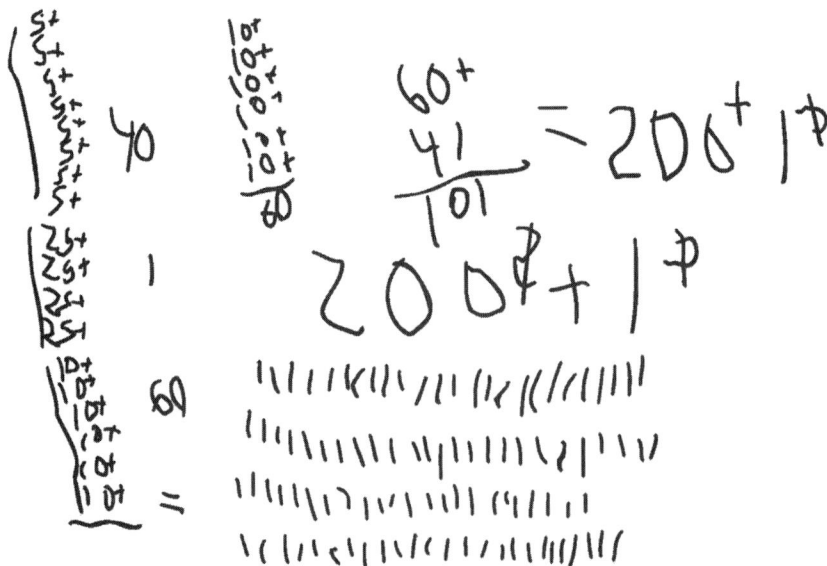

© Developmental Studies Center

"Two hundred plus 1 cent?" I repeated. "Is 200 the same as 2 dollars or is it a different answer?"

Elise seemed uncertain and did not answer. I noticed that Benito, who had reached the answer of 200 through a series of calculations, was shaking his head "no." Thinking this might be an opportunity to incorporate him into the discussion, I asked, "What do you think, Benito?"

He looked uncomfortable, like he wanted to pass. Finally he said, "It's different." When I invited him to show his work and explain his thinking to the class, he shook his head and put his head down on his desk.

I wasn't surprised by Benito's reaction since he was a relatively new student and seemed to have a difficult time connecting with the others. I also noticed that students did not go out of their way to include him. We have a very high turnover rate at my school and students come and go all the time from my class. I try to find ways to help new students become active participants, but I am not always as successful as I would like to be.

Daniel volunteered to show his work next. "For the quarters, I did 25 × 4 equals 100. I did 5 × 8 equals 40 for the nickels. Then I did 6 × 10 equals 60 for the dimes. Then I added the 4 and the 6 and got 10, so the answer is $110."

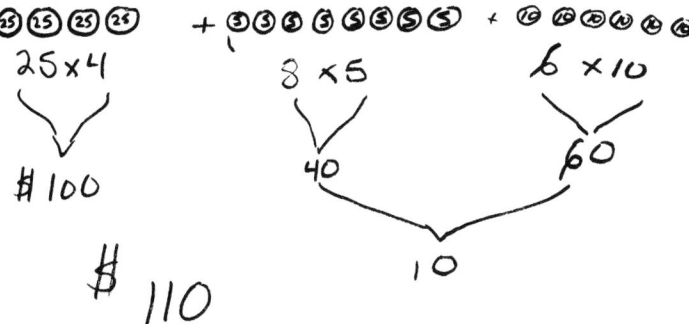

Most of the students in the class had gotten a wrong answer! I could see that they were attempting to apply the strategies they had learned about making tens and multiples of 10. But there was something about the use of dollars and cents that clearly confused them. Some students even forgot the value of the coins, giving each coin a value of "one" despite frequent prior practice with coins. A few students did fine if they were dealing only with dollars or only with cents. Why would thinking about dollars and cents

© Developmental Studies Center

together be so difficult for them? Was this a place value problem, a lack of familiarity with money, or some other difficulty?

Read and Reflect

Bay Area Mathematics Task Force. 1999. *A Mathematics Source Book for Elementary and Middle School Teachers: Key Concepts, Teaching Tips, and Learning Pitfalls.* Novato, CA: Arena Press.

Read pages 23 through 26 on place value concepts and multi-digit computation.

Russell, Susan Jo. 2000. "Developing Computational Fluency With Whole Numbers." *Teaching Children Mathematics 7* (3).

Read pages 154 through 158.

Fuson, Karen, Yolanda De La Cruz, Stephen T. Smith, Ana Maria Lo Cicero, Kristin Hudson, Pilar Ron, and Rebecca Steeby. 2000. "Blending the Best of the Twentieth Century to Achieve a Mathematics Equity Pedagogy in the Twenty-First Century." In M. J. Burke and F. R. Curcio, eds. *Learning Mathematics for a New Century.*

Read pages 197 through 212.

© Developmental Studies Center

Dollars and Cents Confusions
Facilitator's Guide Notes

I n this case, the teacher describes a lesson in which she presents her third- and fourth-grade students with a problem involving money. The problem asks students to find the sum of 4 quarters, 6 dimes, and 8 nickels. The teacher hopes that students will call on their recently acquired mental addition and multiplication computation skills to solve the problem. The case questions assumptions about money as an effective model for students who are just learning the base ten number system, and how well students understand money from their experiences outside of school.

Sample Discussion Issues

What are the benefits and limitations of using money to work with multiples of ten?

What are some alternate ways Benito could participate in lieu of the whole-class discussion?

How do students translate the value of coins into numerical quantities, and what are the likely pitfalls?

Suggested Materials

Play money, including coins and dollar bills

Starter Problem

Solve the following problem. Think about how students might become confused by the different ways to write amounts for dollars and cents.

Elise and Kim went to the school gift shop. They had 4 quarters, 6 dimes, and 8 nickels. How much money did they have together?

Money Confusions

The teacher planned this lesson thinking that money would be a good illustration of the base ten number system for students, and that it would be familiar to them from their experiences outside of school. In her reflections, however, she wonders if students' difficulties with the problem were due to insufficient familiarity with money, inadequate understanding of place value, or some other difficulty.

What can we understand about these students' thinking based on their explanations and their recorded work? Students demonstrated different ways to find the total for a set of coins such as 8 nickels. They added eight 5s, counted by 5s, or multiplied 8 × 5. They also seemed to understand that they needed to add the amounts for the three sets of coins to find how much money Elise and Kim had. The difficulties arose when they needed to make sense of what the amounts 1, 100, 60, and 40 meant.

Elise and Daniel seemed to slip back and forth between thinking of the numbers as pennies and as dollars. In their heads, they sometimes thought of the dollar as 1 and other times thought of it as 100. Similarly, Melinda and Benito were confused about whether 200 was the same as 2 dollars or meant something different. It is easy to confuse what the various numbers and digits mean when working with money. For many students, thinking about these amounts in different ways, and keeping it all straight in their minds, can be difficult.

Multiplication and Repeated Addition

Most students demonstrated that they could use multiplication or repeated addition to solve 25 times 4, 10 times 6, and 5 times 8. Daniel, for example, found the totals for all his coins using multiplication. His error occurred when he tried to combine the totals. Other students, like Rosalie, still relied on counting by 5s to check the total of 8 nickels even when another student suggested that she could multiply 8 times 5 to get 40 cents.

The goal of this lesson is not necessarily to make explicit the relationship between skip-counting (5, 10, 15, 20 . . .), repeated addition (5 + 5 + 5 + 5 + 5 + 5 + 5 + 5 = 40), and multiplication (8 × 5 = 40). The children's strategies, however, provide an excellent starting point for a discussion of this important concept.

Linking Number Sense and the Algorithm

Melinda began by thinking, "Four quarters is $1.00; six dimes makes $1.60." But she was thrown off when she switched from using reasoning to using a paper-and-pencil algorithm. Instead of mentally adding on the nickels, she added $1.60 and 40 and got $5.60, an unreasonable answer. Notice that she didn't put a decimal point in 40 to parallel the decimal point in $1.60. Perhaps she didn't know how to do this, or perhaps she didn't realize that money amounts less than a dollar can be written with a decimal point. When the numbers are written in two different forms, it may make it harder for her to know how to line them up to add. She may wonder, "Where are the tens and ones in $1.60?" Notice that she mistakenly added the numbers by lining them up on the left. Maybe this was because 40 didn't have a decimal point. The errors that Melinda made reveal a schism between the number sense she seems to have about money quantities and her ability to deal with the manipulation of their symbols.

Do Zeros Matter?

Rosalie put "1 dollar and 60 and 40 together and got . . . 11 dollars." It appears that she thought adding 60 and 40 was just like adding 6 and 4 to get 10. Then, when she added 1 more to 10, she got 11. Perhaps the prior lessons, in which students added 3 + 7 and then 30 + 70, led her to believe that these two equations were essentially equivalent, and that the zeros in the answer didn't really matter. After all, students often say that 0 is nothing. Why should it matter if you put them on or take them off?

Student Participation

Concerned about how to help new students become more active participants in her class, the teacher was looking for opportunities to involve Benito. Noticing that Benito was nodding his head at some of the discussion, she thought he might be willing to contribute to it. However, he seemed uncomfortable when asked to share his ideas. Would it have been less intimidating to ask him a question privately and then ask if he would share his response during the whole-class discussion?

Another possibility would be for the teacher to incorporate lower-risk ways for students to participate, such as sharing their ideas with one other person rather than with the whole class. Or perhaps they could write their ideas down, and the teacher could read a few to the class anonymously. As a more comprehensive strategy, the teacher could ask students to brainstorm ways to make new class members feel welcome and help them understand their responsibilities to each other as class members.

Notes

15. Carry Two or Twenty?

Over a period of several weeks, early in the school year, I had presented my third graders with mental computation problems containing numbers close to *benchmark* numbers like 100, 50, 25, and 10. After putting problems like 49 + 99 = ___, 24 × 5 = ___, 363 + 97 = ___, and 545 − 298 = ___ on the board, I would ask students to solve each problem in their heads and, then, turn to a partner to discuss how they solved it. Students then volunteered to report their strategy aloud to the class while I recorded their thinking on the board.

It seemed to me that students were beginning to get comfortable with using these benchmark numbers as referents in doing mental computation. For the problem 49 + 99, for example, I heard explanations like, "I turned *99* into 100. Forty-nine plus 100 is 149, take away 1 is 148."

I noticed during these weeks that some students often volunteered to share their strategies while others never did. I didn't want to call on students unless they volunteered because it seemed like I would be putting them *on the spot.* However, I wanted to hear from everyone at some point, and this clearly was not happening. I also noticed that some students scribbled on little pieces of paper during the problem-solving time. I had told them they could use paper if they needed it to keep track of their thinking. Because they weren't sharing, I had no way of knowing how my quiet students were approaching the problems. I wanted to know if they could use a mental process based on number sense.

Today, I wrote the problem 97 × 4 = ___ on the board and asked students to solve the problem mentally. After about a minute of having them think quietly, I asked them to tell their partner the answer and explain how they solved the problem. As usual, some partners discussed their solution at length while others just said the answer to each other and then sat waiting for me to move on.

© Developmental Studies Center

I then asked students to report their solutions, which I recorded on the board. "Let's start with this one," I said, pointing to *392*. "Let's hear how you solved it."

$$388 \quad 392 \quad 368$$

"Well," Rachel volunteered, "I knew that 90 times 4 is 360, and that 7 times 4 is 28." I recorded what she said on the board.

$$90 \times 4 = 360 \qquad 7 \times 4 = 28$$

"So I had to add 360 and 28," she continued, "And 28 is close to 30 so I did 360 plus 30 and that's 390." She paused as I recorded that.

$$360 + 28 = \underline{\quad} \qquad 360 + 30 = 390$$

"Oops," Rachel said, "The answer is . . . 388."
"What made you change your mind?" I wanted to know.
"I added 2 instead of subtracting 2. That's how I got 392, not 388."
I asked if there were any questions or comments for Rachel and as there were none, we moved on to Julian, who was eagerly waiting, hand in the air.
"Well, I know that 100 times 4 is 400—" he said, as I wrote.

$$100 \times 4 = 400$$

"—and I had to take 12 away from 400."
"Why did you have to do that?" I asked, as I recorded.

$$400 - 12 = \underline{\quad}$$

"Because I added 3 to each 97," he said. "And there were four 97s, so 3 times 4 is 12. I have 12 too much."
"Okay."
"So 400 minus 12 is . . . 400 minus 10 is 390, and minus 2 more is 388." I wrote that on the board.

© Developmental Studies Center

$$400 - 10 = 390 - 2 = 388$$

"Does this strategy make sense?" I asked the class. There was general agreement that it did.

Several more students, hands in the air, were eagerly waiting to share their strategies. I debated whether to stop for a moment to talk about some of the mathematical properties and strategies they were using, but decided to build this into tomorrow's discussion.

David volunteered next. "Well, I did it a different way," he said. "I knew that 7 times 4 is 28."

I recorded.

$$7 \times 4 = 28$$

"No," he corrected me, "You have to record it the other way. Put 97 on top of 4 with the line underneath."

I rewrote the problem vertically, as he requested.

$$\begin{array}{r} 97 \\ \times\ 4 \\ \hline \end{array}$$

He continued, "So 7 times 4 is 28. I put the 8 on the bottom below the 4 and put 2 on the top up by the 9."

I recognized David's strategy as the standard multiplication algorithm. I decided to ask him a question to probe his thinking a bit. I asked, "Where are you getting 2?"

"It's the 2 from the 28."

"If you take 2 from 28, don't you have 26?"

"Oh, 20. Whatever," he said.

"So, you took 20 from 28 and put it up here?" I asked, recording the 20 above the problem.

$$\begin{array}{r} 20 \\ 97 \\ \times\ 4 \\ \hline \end{array}$$

© Developmental Studies Center

"Yeah."

"Then what did you do?"

"I did 4 times 9 and that's 36, then I added 2, I mean 20, and I got . . . 56."

"He means 36 plus 2," Safiya chimed in. "That gives you 38. If you write that down you get 388."

Others in the class agreed, "You have to write it as 2, not 20."

"But David said he meant the 20 from 28," I said.

"Wait, I'm getting confused . . ." David said. He added, rather resentfully, "You didn't record it right."

I pointed to the recording and asked the class, "Does it matter whether we mean 2 or 20 here?"

"You mean 20, but you write 2," Madeline said.

Rachel followed this with: "But she [meaning me!] doesn't want you to do it that way!"

It occurred to me that this was not going very well, especially for David, who had withdrawn into a frustrated silence. Who could blame him?

Bewildered, I decided to leave this and go on to other students who were waiting to share their strategy. Not surprisingly, no one dared bring up the standard algorithm again!

Read and Reflect

Bay Area Mathematics Task Force. 1999. *A Mathematics Source Book for Elementary and Middle School Teachers: Key Concepts, Teaching Tips, and Learning Pitfalls.* Novato, CA: Arena Press.

Read pages 43 through 47 on multiplication concepts.

Caliandro, Christine K. 2000. "Children's Inventions for Multidigit Multiplication and Division." *Teaching Children Mathematics* 6 (6).

Read pages 420 through 423.

© Developmental Studies Center

Carry Two or Twenty?
Facilitator's Guide Notes

In this case the teacher describes a class discussion in which her third-grade students explain their mental computation strategies for solving a multiplication problem. While some students come up with creative solutions based on complex mathematical concepts and properties, others either do not participate or use a standard algorithm to solve the problem. The case raises the issue of how to deal productively with the standard algorithm when the teacher's goal is to encourage solutions based on number sense. It also invites us to think about how we can make full participation in discussions a reality in the classroom.

Sample Discussion Issues

What are the benefits and limitations of asking students to invent their own strategies or algorithms?

What role might *standard algorithms* play in math discussions?

What are some ways that *quiet* students can be given opportunities to participate in mathematical discussions?

Suggested Materials

Base ten blocks and other place value materials

Starter Problem

Solve the following problem mentally. How does your mental computation compare to the standard algorithm? How does it compare to how a third grader might solve the problem?

$$97 \times 4 = \underline{\hspace{1cm}}$$

Mathematical Concepts and Properties

What mathematical concepts and principles did these students draw on to solve the problems mentally? Rachel used the distributive law to break the problem 97 × 4 into (90 × 4) + (7 × 4). She also understood the fundamental place value idea that the 9 in 97 means 90 and that adding 28 is the same as adding 30 and subtracting 2. A name sometimes used for this strategy is *compensation*; adding 2 too many is compensated for by subtracting 2.

Julian solved the same problem by rounding 97 to 100 and multiplying 100 × 4 to get 400. He used a compensation strategy to subtract off the four 3s that were added on to the four 97s resulting in 400 − 12. Instead of subtracting 12 directly, he subtracted 10 and 2 more. He relied on his understanding of place value to know that subtracting 12 was the same as subtracting 10 and 2. Both Rachel and Julian used *anchor numbers,* such as multiples of 10, to make their computations simple.

David, who used the standard algorithm, might have known that although he was carrying 2, it meant 20, even though he became confused when the teacher wrote 20 above the 9 in 97. This might have been an opportunity to help students review the role of place value in the algorithm and talk about what 20 would mean if it were placed above the 9 in 97.

Language Confusions: Place Value Language

As teachers, we may use place value language loosely and actually provide an incorrect model for students. Do we say *carry two* when we mean *carry two tens?* Do we say there are 2 tens in 320 when there are actually 32 tens? What we mean is that there is a 2 in the tens place. Confusion can result when we inadvertently misuse place value language in these ways. Careful attention to language may have prevented the confusion that arose when the teacher wrote 20 over the 9 in 97. This may prompt students to think they are *carrying* 20 tens, which is incorrect.

Developing Flexibility and Efficiency

What is the role of standard algorithms in modern mathematics education? One school of thought is that, with calculators so readily available, standard algorithms are becoming obsolete. Others believe that standard algorithms are counterintuitive and that learning them is actually harmful to children's development of number sense. Others believe that mastery of standard algorithms through repeated practice is the foundation of reliable computation and is necessary for success on standardized tests.

One compromise is to encourage students to develop computational strategies based on reasoning, while not discouraging them from also employing the traditional algorithm. In this scenario, students could discuss which strategies are more efficient and have fewer pitfalls in different situations and for different numbers. They would realize, for example, that it is almost silly to use an algorithm for a problem like 300 − 299. On the other hand, they might choose to use a standard algorithm for the problem 563 − 384, even though they could use reasoning to solve it. For many students, the standard algorithm is an efficient, comfortable, and reliable routine. However, students who are wedded to the algorithm may not fully develop, or may fail to draw on, their number sense.

A Safe Environment for Risk-Taking

What was expected to be an engaging discussion disappointingly yielded confusion and hurt feelings. Thankfully, students like Safiya and Madeline seemed accustomed to listening to each other and chiming in to help their classmates along. What happened that made the classroom environment suddenly feel *unsafe* for David? And why did this situation not arise when students were sharing other strategies?

Perhaps David felt singled out because the teacher had seemed open to all strategies, choosing to play devil's advocate only with his. It might be helpful for the teacher to adopt a habit of discussing the benefits and drawbacks of all strategies that students bring up. Part of building a safe environment that also encourages risk-taking is assuring students that all ideas will be given sincere consideration and letting them know that challenging questions will be asked about many ideas, not just about one student's idea.

Although some students were demonstrating very sophisticated mental strategies, others were not sharing strategies. Perhaps their mathematical understanding and confidence to speak would have been bolstered had the mathematical ideas and principles been discussed more explicitly rather than just sharing strategies. It also may have been important to hear other students explain how a strategy *popped into their heads*. For example, it might have helped if Rachel had explained why she decided to start with 90×4 or if Julian had explained why he decided to take 12 away from 400. Additionally, the teacher might have asked students to turn to their partners to discuss their strategies or to discuss a question prior to discussion with the whole class. Partner talk increases students' accountability for participation and offers a safer environment for students who are still uncomfortable speaking to a larger group.

Notes

16. How Many Can She Buy?

At my school most of the teachers ask students to solve problems and discuss their strategies as a way to learn math. Our students have responded very positively to this way of doing math and their progress reflects what might be expected in higher-income communities.

In mid-October I presented a group of fourth graders with a multistep mathematics problem. I thought the first part of the problem would be easy, and that we would spend most of our time working on and discussing the second part.

Prior to this, we had focused on mental computation problems using addition and multiplication. The students seemed comfortable solving problems using rounding to a convenient number (for example, solving $2 \times 96 = $ ___ by thinking $2 \times 100 = 200, 200 - 8 = 192$), partitioning (solving $12 \times 7 = $ ___ by thinking 10×7 plus 2×7), and doubling (solving $4 \times 17 = $ ___ by thinking 17 doubled is 34, 34 doubled is 68).

I handed out the problem to students and asked them to follow along as I read it aloud:

> Mrs. Kane has 97 cents in her pocket. Chocolate candies cost 4 cents each.
> Part 1: How many chocolate candies can she buy?
> > Does she have enough chocolate candies to give one to each student in our class?
> Part 2: If she does, how much money does she have left?
> > If she doesn't, how much more money does she need?

Initially, many students latched on to "give one to each student in our class" and began counting their classmates. When I asked a couple of students why they were counting, they responded, "To see if there are enough."

"Enough of what?" I asked.

© Developmental Studies Center

"Enough chocolate candies," they answered.

Yet, when I asked if they knew how many chocolate candies there were, at least one student told me "97."

I asked them to tell me what the problem was asking. After rereading the problem, they seemed to understand and started working. I walked around the room to observe how they were approaching the problem and to ask occasional questions about their work. I noticed that many students, like Sarah for example, seemed to be adding a lot of fours on their papers. Also, like Sarah, many were making computational errors.

With about fifteen minutes remaining, I brought the class together even though at least ten students had not solved the first part of the problem, and almost no one had solved the second part. "Let's hear what you got for the first part of the problem," I said.

"I got 24," Midori said. I wrote 24 on the chart paper. Most students expressed agreement with Midori's answer, and I wondered if they were agreeing because 24 seemed like a reasonable answer, or because her peers perceived her as a smart math student.

Jason disagreed, however, and told the class his answer was 23. The class responded with a chorus of "Disagree."

I recorded Jason's answer next to Midori's without comment and told the class that it was up to them to prove whether 23 or 24 was the correct answer.

Khlana raised her hand. "I think 24 is correct," she said. "I counted by 4s up to 96." As Khlana counted aloud, I recorded each number on chart paper, "4, 8, 12, 16 . . ." so that the class could follow her thinking. Then she said, "I went back and counted how many numbers I wrote down, and it was 24" (see the figure on p. 178). Khlana went on to say that she could also write an equation, $4 \times 24 = 96$. Then she explained, "I had 24 fours, that's 24 chocolate candies. It's 96 cents all together."

The class agreed with Khlana's strategy and equation, so I went back to Jason, who had given the answer of 23. I thought this might be a good opportunity for the class to monitor an incorrect procedure.

"I used take-away," he began. "I started at 97 and took away 4. That was 93."

I recorded his thinking on the board as he read his equations aloud: "$97 - 4 = 93, 93 - 4 = 89, 89 - 4 = 85, 85 - 4 = 81, 81 - 4 = 77, 77 - 4 = 73, 73 - 4 = 69, 69 - 4 = 65, 65 - 4 = 61, 61 - 4 = 57, 57 - 4 = 53, 53 - 4 = 47.$"

© Developmental Studies Center

NAME: Sarah

Mrs. Kane has 97 cents in her pocket. Chocolate candies cost 4 cents each.

Part 1: How many chocolate candies can she buy? Does she have enough chocolate candies to give one to each student in our class?

24 is as much as Mrs. Kane Can buy!

Part 2: If she does, how much money does she have left? If she doesn't, how much more money does she need?

Show your work below and then record your answers above.

Part 1:

```
  4          54
+ 4        + 4
  8          58
+ 4         +4
 12          62
+ 4         + 4
 16          64
+ 4         + 4
 20          68
+ 4         +4
 24
+ 4          72
 28          +4
+ 4          74
 32          + 4
+ 4          78
 38          +4
+ 4          82
 42          +4
+ 4          86
 46          + 4
+ 4          90
 50          + 4
+ 4          94
 54          + 4
             98
             - 1
             97
```

Part 2:

© Developmental Studies Center

NAME: Khlana

Mrs. Kane has 97 cents in her pocket. Chocolate candies cost 4 cents each.

Part 1: How many chocolate candies can she buy? Does she have enough chocolate candies to give one to each student in our class?

She can buy 24. She needs to buy 8 more.

Part 2: If she does, how much money does she have left? If she doesn't, how much more money does she need?

Show your work below and then record your answers above.

4, 8, 12, 16, 20, 24, 28, 32, 36, 40, 44, 48, 52, 56, 60, 64, 68, 72, 76, 80, 84, 88, 92, 96

I counted by 4's

I also did 4 × 24 = 96

At this point, the class was quickly losing interest in Jason's strategy since it was rather inefficient and time-intensive. I wish I had a better way to keep the interest of the class through tedious explanations such as this. I need to have them listen to their classmates since I am relying on them to detect each other's errors and learn from each other.

One or two students who were still paying attention said, "Disagree," and one of them added, "It's 49, not 47."

Jason stopped and looked at his work for a long moment while the class fidgeted and got off-task. Finally he said, "Oh, yeah! It's 49. I made a mistake."

© Developmental Studies Center

NAME: Antonio

Mrs. Kane has 97 cents in her pocket. Chocolate candies cost 4 cents each.

Part 1: How many chocolate candies can she buy? Does she have enough
chocolate candies to give one to each student in our class?

She can get 24 chocolate candies.

Part 2: If she does, how much money does she have left? If she doesn't, how
much more money does she need?

She has 1 cent left

Show your work below and then record your answers above.

1. 24 × 4 = 96   Mrs. Kane can buy 24
chocolate candies.

20 × 4¢ = 80   4 × 4¢ = 16   16 + 80 = 96

2. She has 1 cent left because
she bought 24 chocolate
candies with 96 cents.
She can't give every
student in the class one
because she can only
buy 24 with 96 cents
and she only had 97
cents in her pocket.

© Developmental Studies Center

He continued to subtract 4s until he got to one. Then he added, "I went back and counted the 4s, and it was 23,"

"Do you still think it will be 23?" I asked.

"I'm not sure," he replied.

I thought I could regain the attention of the class by suggesting that we count the fours together from my recording of Jason's strategy. When the class landed on 24, Jason said, "I want to change my answer to 24."

The next student to volunteer to share his solution was Antonio, a very capable mathematics student. He was one of the few who had completed both parts of the problem, as shown in the figure on p. 179.

"Put $20 \times 4 = 80$." He began, and I recorded his equation on the board. "Then I put $4 \times 4 = 16$, and $16 + 80 = 96$, so that's 24 chocolate candies."

"How did you decide to use 20×4 and 4×4, and how did you know how many chocolate candies there were?"

"I knew $4 \times 20 = 80$, and that is close. Then I added on 4s until I reached 96. That's four more 4s or 4×4. I added that to 20 and got 24 chocolate candies." Unfortunately, the bell rang before the conversation could be carried any further.

So what do we do tomorrow? I saved the chart paper from our discussion so that we could go back to the strategies presented today. I was a little disappointed that only a couple of students used the mental math strategies that we had been studying. The same two students also used multiplication. No one used division, even though they had been introduced to division in third grade. Perhaps tomorrow I can help them reexamine their strategies and discover why some operations can be shortcuts for other operations.

Read and Reflect

Bay Area Mathematics Task Force. 1999. *A Mathematics Source Book for Elementary and Middle School Teachers: Key Concepts, Teaching Tips, and Learning Pitfalls.* Novato, CA: Arena Press.

Read pages 37 through 41 and 48 through 54 on multiplication and division concepts.

Russell, Susan Jo. 2000. "Developing Computational Fluency With Whole Numbers." *Teaching Children Mathematics* 7 (3).

Read pages 154 through 158.

© Developmental Studies Center

How Many Can She Buy?
Facilitator's Guide Notes

This case describes a lesson in which students share strategies for solving a multistep problem with their classmates. Disappointingly, the majority of the students use strategies such as counting, repeated addition, and repeated subtraction, rather than more efficient ones. During the whole-class sharing, students lose interest in their classmates' tedious explanations. This concerns the teacher because they are expected to actively listen to and verify each other's strategies. The teacher considers what to do the following day to help students use more efficient mental math strategies and to explore how some operations (multiplication and division) are shortcuts for other operations (repeated addition and repeated subtraction).

Sample Discussion Issues

Why might the problem prompt students to use repeated addition instead of multiplication?

How do multiplication and division work as shortcuts for repeated addition and repeated subtraction?

What are the benefits and drawbacks of putting an incorrect solution on the board for discussion?

Suggested Materials

Play money, base ten materials

Starter Problem

Solve the following problem four different ways, by using addition, subtraction, multiplication, and division.

Mrs. Kane has 97 cents in her pocket. Chocolate candies cost 4 cents each. How many chocolate candies can she buy?

Multiplication and Division for Efficiency

The students had learned several mental computation strategies previously, namely rounding, partitioning, and doubling. Yet we know of only one student who drew on this knowledge. Antonio multiplied a number by 4 to get a total close to 97. It is interesting that neither he, nor other students, appeared to use knowledge about the proximity of 97 to 100, a number easily divisible by 4, in their solution. One thing the teacher might do the following day is to ask students to reexamine Antonio's strategy and think about how knowing how many 4s in 100 might help them know how many 4s are in 97. Since 4×25 is 100, then the missing number for $4 \times \underline{\quad} = 97$ would be a little less than 25.

Khlana counted by 4s to get to 96 and got twenty-four 4s. Then she wrote a multiplication equation to reflect her solution, $4 \times 24 = 96$. Most other students, such as Sarah, also used counting or repeated addition to solve the problem. Why didn't they use multiplication, a shortcut for repeated addition? Perhaps the reason is that the equation needed to solve the problem is $4 \times \underline{\quad} = 96$.

In other words, the students know that each candy costs 4 cents, but they don't know what number to multiply by. Notice that Khlana wrote her equation after she figured out the answer, not to solve the problem. If students could frame the problem as a missing multiplier or missing multiplicand, they might be able to use mental computation with multiples of ten to figure out the missing number, as Antonio did.

Jason used repeated subtraction to find out how many 4s he could take away from 97. He made a mistake when subtracting one of the numbers and got the incorrect answer. Although his strategy is correct, it is also prone to computation errors because it involves so many steps. Perhaps it would be helpful for students to examine the relationship between Jason's method of taking away 4s, one at a time, and dividing by 4, where you can *take away* several 4s at a time. So, for example, to divide 97 by four they might ask themselves, "How many 4s are in 90?" They know that since 20×4 is 80, there are twenty 4s in 90 with 17 left. Then they could ask, "How many 4s are in 17?" They would find that there are four 4s with 1 left. That's twenty-four 4s in 97 with 1 left over. The relationship between repeated subtraction and division is similar to the relationship between repeated addition and multiplication. Division and multiplication might be thought of as *shortcuts*.

Increasing Participation and Active Listening

The teacher decides to proceed with a whole-class sharing of strategies even though at least ten students have not solved the first part of the problem, and most students have not solved the second part. During the longer, more tedious explanations, and particularly in an erroneous one, the students lose interest and get off task. How can the teacher help students take responsibility for listening to, understanding, verifying, and learning from each other's strategies?

One idea is for the teacher to occasionally intersperse questions during the sharing time for partners to discuss. For example, the teacher might increase class participation by asking pairs of students to talk about a shortcut for taking away a lot of 4s at one time. Students might also be asked to reflect on the discussion process afterwards and give a suggestion for what they could do to demonstrate to their peers that they are listening actively during sharing time.

Agreeing and Disagreeing Responsibly

The class responds with *agree* or *disagree* when students share their strategies aloud. What might be the purpose of this way of responding? By asking for group responses, the teacher is providing ways for more students to participate. Inviting disagreement also sets the stage for discussing, and learning from, other students' errors. Antonio, for example, seemed comfortable presenting his solution, even though his answer was in the minority. He also was open to changing his mind about his answer when it was proven incorrect.

At one point, the teacher wonders if students are agreeing with Midori's answer because they believe it is reasonable, or if it is because of Midori's reputation as a smart math student. What alternatives might there be to this teachers' approach for holding students accountable?

Bibliography

Barnett, Carne, Donna Goldenstein, and Babette Jackson (eds.). 1994. *Teaching Mathematics Cases, Fractions, Decimals, Ratios and Percents: Hard to Teach and Hard to Learn?* Portsmouth, NH: Heinemann.

Bay Area Mathematics Task Force. 1999. *A Mathematics Source Book for Elementary and Middle School Teachers: Key Concepts, Teaching Tips, and Learning Pitfalls.* Novato, CA: Arena Press.

Beishuizen, Meindert. 1993. "Mental Strategies and Materials or Models for Addition and Subtraction Up to 100 in Dutch Second Grades." *Journal of Research in Mathematics Education* 24 (4): 294–323.

Bishop, Alan J. 2001. "What Values Do You Teach When You Teach Mathematics?" *Teaching Children Mathematics* 7 (6): 346–349.

Caliandro, Christine K. 2000. "Children's Inventions for Multidigit Multiplication and Division." *Teaching Children Mathematics* 6 (6): 420–423.

Carroll, William M., and Denise Porter. 1997. "Invented Strategies Can Develop Meaningful Mathematical Procedures." *Teaching Children Mathematics* 4 (7): 370–374.

Chambers, Donald. 1996. "Direct Modeling and Invented Procedures: Building on Student's Informal Strategies." *Teaching Children Mathematics* 3 (2): 92–95.

Clements, Douglas H. 1997. "(Mis?)Constructing Constructivism." *Teaching Children Mathematics* 4 (4): 198–200.

Clements, Douglas H. 1999. "Subitizing: What Is It? Why Teach It?" *Teaching Children Mathematics* 5 (7): 400–405.

Cotter, Joan. 2000. "Using Language and Visualization to Teach Place Value." *Teaching Children Mathematics* 7 (2): 108–114.

Curcio, Francis R., and Sydney L. Schwartz. 1998. "There Are No Algorithms for Teaching Algorithms." *Teaching Children Mathematics* 5 (1): 26–30.

Fuson, Karen, Yolanda De La Cruz, Stephen T. Smith, Ana Maria Lo Cicero, Kristin Hudson, Pilar Ron, and Rebecca Steeby. 2000. "Blending the Best of the Twentieth Century to Achieve a Mathematics Equity Pedagogy in the Twenty-First Century." In Burke and Curcio (eds.), 197–212 *Learning Mathematics for a New Century.* Reston, VA: NCTM.

Fuson, Karen C., Laura Grandau, and Patricia A. Sugiyama. 2001. "Achievable Numerical Understandings for All Young Children." *Teaching Children Mathematics 7* (9): 522–526.

Geary, David. 1998. "Developing Arithmetical Skills." *Children's Mathematical Development.* Washington, DC: American Psychological Association.

Geary, David. 1998. "Learning Mathematical Problem Solving." *Children's Mathematical Development.* Washington, DC: American Psychological Association.

Isaacs, Andrew C., and William M. Carroll. 1999. "Strategies for Basic-Facts Instruction." *Teaching Children Mathematics 5* (9): 508–515.

Kamii, Constance, Barbara A. Lewis, and Bobbye M. Booker. 1998. "Instead of Teaching Missing Addends." *Teaching Children Mathematics 4* (8): 458–461.

Khisty, Lena Licón. 1995. "Making Inequality: Issues of Language and Meanings in Mathematics Teaching with Hispanic Students." In *New Directions for Equity in Mathematics Education,* Walter G. Seceda, Elizabeth Fennema, and Lisa Byrd Adajian (eds.), 279–297. New York: Cambridge University Press.

Klein, Anton S., Meindert Beishuizen, and Andri Treffers. 1998. "The Empty Number Line in Dutch Second Grades: Realistic Versus Gradual Program Design." *Journal for Research in Mathematics Education* 29 (4): 443–464.

Kline, Kate. 1998. "Kindergarten Is More Than Counting." *Teaching Children Mathematics 5* (2): 84–87.

MacGregor, Mollie, and Kaye Stacey. 1999. "A Flying Start to Algebra." *Teaching Children Mathematics 6* (2): 78–85.

National Council of Teacher of Mathematics. 2000. *Principles and Standards for School Mathematics,* 17–19. Portland, OR: Graphics Arts Center.

Russell, Susan Jo. 2000. "Developing Computational Fluency with Whole Numbers." *Teaching Children Mathematics 7* (3): 154–158.

Trafton, Paul R., and Carol Midgett. 2001. "Learning Through Problems: A Powerful Approach to Teaching Mathematics." *Teaching Children Mathematics 7* (9): 532–536.

Witherspoon, Mary Lou. 1999. "And the Answer Is . . . Symbolic Literacy." *Teaching Children Mathematics 5* (7): 396–399.

—— ALSO AVAILABLE ——

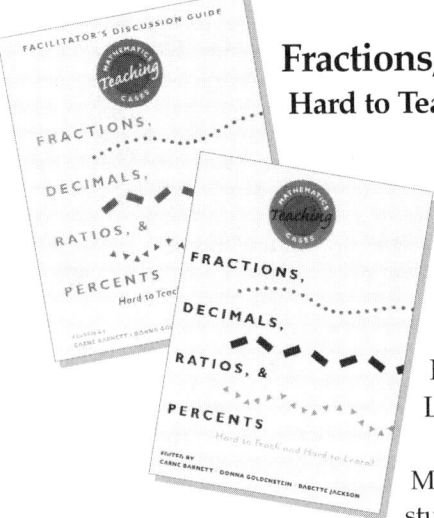

Fractions, Decimals, Ratios, & Percents
Hard to Teach and Hard to Learn?

EDITED BY
Carne Barnett, WestEd
Donna Goldenstein and
Babette Jackson

FOREWORD BY Judith H. Shulman and
Lee S. Shulman

Math teachers know all too well that when students enter the abyss of fractions, decimals, ratios, and percents, problems ensue. Students who have previously succeeded begin to falter, their confidence wanes, failures increase, and teachers despair.

Barnett, Goldenstein, and Jackson confront this problem by using one of the most effective and underused resources educators possess—their stories, insights, and experiences. Written by upper elementary and middle school teachers, the cases detail their experiences teaching fractions, decimals, ratios, and percents, exemplifying the recurring obstacles most teachers face. The typical case reports a problem, dilemma or crisis of mathematics instruction. It describes the events that led up to the problem, the classroom context in which it occurred, and the ways in which the teacher attempted to resolve it.

The *Facilitator's Discussion Guide* provides both beginning and experienced teachers with practical suggestions for initiating and leading case discussion groups. Extensive teaching notes for each case help teachers anticipate issues. Together the cases and guide will expand teachers' mathematical knowledge, challenge beliefs, and encourage innovative changes in teaching.

Casebook
0-435-08357-0 / 127pp

Facilitator's Discussion Guide
0-435-08358-9 / 93pp

To save 10% or for more infomation, visit us online: www.heinemann.com